SCIENCE
THROUGH
ACTIVE
READING

Teacher's Edition

Life Science

Content and Learning Strategies

Teacher's Edition

Life Science
Content and Learning Strategies

Sharron Bassano
Mary Ann Christison

Addison-Wesley Publishing Company

Reading, Massachusetts • Menlo Park, California • New York
Don Mills, Ontario • Wokingham, England • Amsterdam
Bonn • Sydney • Singapore • Tokyo • Madrid • San Juan

Life Science: Content and Learning Strategies. Teacher's Edition

Copyright © 1992 by Addison-Wesley Publishing Company.
All rights reserved.
No part of this publication may be reproduced,
stored in a retrieval system, or transmitted
in any form or by any means, electronic, mechanical,
photocopying, recording, or otherwise,
without the prior permission of the publisher.

Executive editor: Joanne Dresner
Development editor: Debbie Sistino
Production editor: Janice L. Baillie
Text design: Joseph DePinho
Cover design: Joseph DePinho
Cover photo: François Gohier/Photo Researchers
Text art: Lloyd P. Birmingham

ISBN: 0-8013-0985-9

2 3 4 5 6 7 8 9 10-AL-96959493

CONTENTS

To the Teacher — ix
Answer Key — xv
To the Student — xxi

CHAPTER 1

BOTANY Angiosperms: The Largest Class of Plants — 1

CRITICAL THINKING ACTIVITIES — 2

CONTENT READINGS
  1 ★ Characteristics of Angiosperms — 4
  2 ★ The Leaf — 8
  3 ★ The Roots — 11
  4 ★ The Stem — 13
  5 ★ The Flower — 16

LEARNING STRATEGIES
Using prior knowledge ☆ Working cooperatively ☆ Imagery ☆ Taking notes ☆ Self-evaluation ☆ Reading selectively ☆ Grouping

GLOSSARY — 22

CHAPTER 2	ZOOLOGY Mammals: Animals We Know Best	23
	CRITICAL THINKING ACTIVITIES	24
	CONTENT READINGS	
	1 ★ Differences in Animals	27
	2 ★ Characteristics of Mammals	30
	3 ★ Habitats of Mammals	34

> **LEARNING STRATEGIES**
>
> Using prior knowledge ☆ Working cooperatively ☆ Inferencing ☆ Self-evaluation ☆ Reading selectively ☆ Using resources ☆ Grouping

GLOSSARY 38

CHAPTER 3	HUMAN ANATOMY Skin, Muscles, and Bones of the Human Body	40
	CRITICAL THINKING ACTIVITIES	41
	CONTENT READINGS	
	1 ★ The Skin	44
	2 ★ The Muscles	47
	3 ★ The Skeletal System	51

> **LEARNING STRATEGIES**
>
> Using prior knowledge ☆ Working cooperatively ☆ Inferencing ☆ Self-evaluation ☆ Reading selectively ☆ Taking notes

GLOSSARY 56

CHAPTER 4	HUMAN PHYSIOLOGY Digestion, Respiration, and Circulation	58
	CRITICAL THINKING ACTIVITIES	59
	CONTENT READINGS	
	1 ★ Digestion	62
	2 ★ Respiration	65
	3 ★ Circulation	68

> **LEARNING STRATEGIES**
>
> Using prior knowledge ☆ Working cooperatively ☆ Inferencing ☆ Self-evaluation ☆ Reading selectively ☆ Grouping

GLOSSARY 74

CHAPTER 5

HUMAN PHYSIOLOGY The Sense Organs — 75

    CRITICAL THINKING ACTIVITIES — 76

    CONTENT READINGS
- 1 ★ The Sense of Hearing — 78
- 2 ★ The Senses of Taste and Smell — 81
- 3 ★ The Sense of Sight — 84
- 4 ★ The Sense of Touch — 87

> **LEARNING STRATEGIES**
>
> Using prior knowledge ☆ Working cooperatively ☆ Inferencing ☆ Self-evaluation ☆ Taking notes ☆ Reading selectively ☆ Imagery

    GLOSSARY — 90

CHAPTER 6

HUMAN ECOLOGY Healthful Living: Nutrition and Exercise — 92

    CRITICAL THINKING ACTIVITIES — 93

    CONTENT READINGS
- 1 ★ Nutrients and Healthful Eating — 95
- 2 ★ The Basic Food Groups — 97
- 3 ★ Exercise — 100

> **LEARNING STRATEGIES**
>
> Using prior knowledge ☆ Working cooperatively ☆ Inferencing ☆ Self-evaluation ☆ Taking notes ☆ Reading selectively ☆ Grouping

    GLOSSARY — 103

TO THE TEACHER

WHO IS THIS BOOK FOR?

Life Science: Content and Learning Strategies is for middle school and secondary ESL students or students experiencing difficulty with a traditional science textbook. *Life Science* is designed to help students develop the science vocabulary, critical thinking skills, and learning strategies needed to pursue higher level content-area schoolwork.

In order to benefit from this text, an ESL student should:

★ be at an intermediate fluency stage of language acquisition
★ have already acquired basic, interpersonal skills in English
★ read and write in English at a simple sentence level

WHERE CAN THIS BOOK BE USED?

Life Science can be used in a variety of instructional settings. It is especially suited for:

★ Sheltered English classrooms
★ content-area resource centers
★ mainstream science classrooms

An ESL teacher who may have little formal background in life science will benefit from regular conferences with the mainstream science instructor. A science teacher with little formal background in language instruction will benefit from regular conferences with the ESL teacher.

All teachers of content-area materials will benefit from instruction that integrates academic vocabulary and definitions, content-specific concepts, and learning strategies. A note about using learning strategies, as well as a list of learning strategies featured throughout this text, appears on pages xii and xiii.

WHAT IS THE GOAL OF THIS BOOK?

Life Science uses a cooperative learning environment to prepare limited English proficient students for mainstream life science classes. By integrating content-area reading comprehension, vocabulary, and learning strategies with hands-on science activities, *Life Science* aims to elevate students' levels of experience, ability, and self-confidence so they can move successfully into higher level courses.

HOW IS THIS BOOK DIFFERENT FROM MAINSTREAM SCIENCE TEXTBOOKS?

Life Science is an introductory overview of the field of life science. It emphasizes active reading comprehension and learning strategies that will enable students to successfully prepare for a mainstream life science textbook. It does this through:

- ★ a consistent, easy-to-follow format with step-by-step instructions to the student
- ★ an abundance of clear, easy-to-understand illustrations used to clarify vocabulary and concepts
- ★ a strong emphasis on oral language activities in cooperative groups or pairs that encourage peer-tutoring and sharing of information
- ★ a variety of critical thinking activities designed to motivate students' interest in the content area, to draw on their prior knowledge, and to make them aware of learning strategies they are employing
- ★ vocabulary clarification activities that require students to use content clues, glossary, and illustrations to improve their understanding
- ★ self-evaluation and extension activities that maximize student opportunities for discussion and review, and encourage the development of student self-confidence

HOW IS EACH CHAPTER ORGANIZED?

Introduction. Each chapter begins with a very brief introduction to be read by the teacher and students together. This introduction puts the chapter topic in a realistic context and provides a focus.

Critical Thinking Activities. Before any readings, a variety of activities provide an overview of the chapter. Through group discussion and experiments, students exercise vital critical thinking skills in advance of the readings. By predicting, hypothesizing, comparing and contrasting information, students are motivated to learn more about the content area. These critical thinking activities are crucial to student success, allowing them to manipulate vocabulary and concepts that will help make the text comprehensible.

What Do You Already Know about the Topic? This section personalizes what students already know about the topic and reviews general-knowledge English vocabulary. It is also the initial peer-tutoring experience. When students complete the task, the whole group reviews any questions about meaning or pronunciation.

Think about These Ideas. This section offers small-group critical thinking activities that build predicting skills, and ask students to draw on previous knowledge or use their imaginations. Students are encouraged to learn from one another. When the groups have finished, the class as a whole talks about their ideas or guesses. After reading the chapter they will come back to these questions and see if they have changed their ideas. This is the time for students to hypothesize, and not to necessarily get the "one right answer."

Group Observations. These tasks will require some materials or props. They are hands-on experiments carried out in groups. Again, reaffirm to students that this is a time to look, listen, think, take notes, and assist one another. Group observations deal with concepts that will be clarified in the readings to come. They are intended to stimulate questions in the minds of the students and motivate them to want to read.

Pre-Reading/Readings. Each of the readings per chapter is prefaced with a pre-reading set of **focus** and **detail questions** that instruct students to skim and scan for specific information. These tasks allow the students to practice the learning strategies of advance organization and reading selectively, so that they seek out the most important words and ideas from the reading. Students locate the key concepts and take notes. They learn to notice new vocabulary, and how to use context or the end-of-chapter glossary to define new words. These and other skills and strategies develop simultaneously through reading the material two or more times. Answers to questions are found in the answer key on pages xv–xx of this teacher's edition.

> *A Word about Vocabulary Tickets.* In advance of reading, students should be given a small handful of **vocabulary tickets**—small slips of paper on which to jot down any word or phrase they are not sure about. Students drop the tickets into a central box or basket after finishing the reading. These tickets now belong to the group, and will be clarified by and for the whole group. This activity assures individual students that their questions will be answered in a reassuringly anonymous manner. Using vocabulary tickets lowers stress and encourages students to ask for help, which they are often reluctant to do.

Self-Evaluation. At the end of each reading, three cooperative learning activities—*Vocabulary Tickets, Vocabulary Check,* and *Question Review*—are suggested for vocabulary clarification, pronunciation practice, and review of concepts. These three activities are intended to provide closure through peer support. They confirm to each student that he or she has been able to extract and evaluate essential data from the reading and has been able to assist someone else, if necessary.

Extension Activities. After all the readings in each chapter, extension activities provide an opportunity for students to look again at their initial hypotheses and to apply what they have learned to new situations.

Glossary. Each chapter has a glossary that includes the content-specific vocabulary printed in boldface type within the reading. Students should know that these heavy-print words are glossed at the end of each chapter and that these are the same words found in the vocabulary check activity at the end of each reading. It may also be helpful to do several glossary activities in class to acquaint students with this resourcing strategy for understanding new words. Use of the glossary can lead to dictionary activities through which more advanced students can discover other kinds of information about these or other words from the reading.

A NOTE ABOUT LEARNING STRATEGIES

Successful students use a variety of techniques, or strategies, to help them learn new academic skills. For example, a good reader uses the strategy of **advance organization** to preview the main ideas of a text, and then **reads selectively** to locate specific facts or ideas. If illustrations accompany the text, the reader will use **imagery** to help unlock meaning. If unfamiliar vocabulary or concepts are presented in the text, the skillful reader will use **inferencing** and context clues.

Throughout a reading experience, the successful learner will make use of **prior knowledge**, or using what the reader already knows, to facilitate learning new information about a subject. The learner will then use such strategies as **taking notes** and **imagery** to keep the experience fresh and make it available for future reference.

Learning strategies are used not only by students. They are also used by successful listeners, speakers, and writers. They apply to all modes of learning and to all content areas. *Life Science* gives students repeated opportunities to use multiple learning strategies so they will apply the strategies to new learning situations. The key strategies are identified in the left-hand margin of each section, but many additional strategies are also found in the lesson. For example, **working cooperatively** and **questioning for clarification** are embedded in virtually every task. Students are **using resources** (the glossary) and the strategy of **advance organization** (focus questions) in every reading. Your role as teacher is to help students choose the appropriate strategy and then to encourage them to become aware of their learning techniques, and to apply learning strategies to different learning activities.

Three types of learning strategies are present in this text:

- ★ **Metacognitive strategies.** These involve higher order thought processes in planning for learning, monitoring the learning while it is taking place, and evaluating this learning.
- ★ **Cognitive strategies.** These are strategies through which the learner interacts with the material being learned. He or she does so by manipulating the material mentally or physically. Cognitive strategies include making mental or actual images ("imagery"), taking notes, grouping items to be learned in meaningful categories, inferring a meaning from context, and building on prior knowledge.
- ★ **Social-affective strategies.** These are strategies that involve either interacting with another person or persons to assist learning or using some kind of affective or feeling control to assist

a learning task. Social-affective strategies include working cooperatively and asking questions for clarification.

LEARNING STRATEGIES* USED IN *LIFE SCIENCE*

Metacognitive Strategies

Advance organization Previewing the main ideas and concepts of the material to be learned

Reading selectively Attending to, or scanning key words, phrases, linguistic markers, sentences, or types of information

Self-evaluation Judging how well one has accomplished a learning activity after it has been completed

Cognitive Strategies

Using prior knowledge Using what is already known to facilitate a learning task

Grouping Classifying words, terminology, or concepts according to their attributes

Imagery Creating or using visual images (either mental or actual) to understand and remember new information

Inferencing Using information in the text to guess meanings of new items, predict outcomes, or complete missing parts

Taking notes Writing down key words and concepts in abbreviated form during a listening or reading activity

Using resources Using reference materials such as glossaries, dictionaries, encyclopedias, and textbooks

Sequencing Organizing information according to a logical order

Interpreting data Using information gathered from maps, graphs, tables, and charts

Social-Affective Strategies

Working cooperatively Working together with peers to solve a problem, pool information, check a learning task, or get feedback on oral or written performance

Questioning for clarification Eliciting from a teacher or peer additional explanation, rephrasing, or examples

---

*Chamot, A. U., and O'Malley, J. M. (1986). *A Cognitive Academic Language Learning Approach: An ESL Content-Based Curriculum.* Wheaton, MD: National Clearinghouse for Bilingual Education.

ANSWER KEY

| CHAPTER | BOTANY Angiosperms: The Largest Class of Plants |

PRE-READING 1

Focus Question

Angiosperms are the largest class of plants. They are green plants with seeds and flowers.

Detail Questions

1. Roots, stems, seeds, flowers
2. The transport system carries food and water through the plant.
3. Monocots and dicots
4. The seed, the stem, the roots, and the transport system

PRE-READING 2

Focus Question

The leaf carries out photosynthesis, making food for the plant.

Detail Questions

1. A frond is a large, flat leaf.
2. Sunlight and carbon dioxide for the plant
3. A leaf breathes in carbon dioxide.
4. Photosynthesis occurs.
5. It breathes out oxygen.
6. Chlorophyll
7. Energy from the sun.

PRE-READING 3

Focus Question

Roots pull water from the soil, hold the plant firmly in the ground, hold soil around the plant, and store food and water for the plant.

Detail Questions

1. The epidermis protects the root and pulls in water. The cortex stores water and food, such as sugar. The vascular cylinder moves the water and food through the plant.

2. A taproot is a flat, thick root.
3. A taproot
4. Fibrous roots look like long, thin strings.
5. Fibrous roots

PRE-READING 4

Focus Question

Stems hold the leaves up to the sunlight, carry sugar from the leaves to the rest of the plant, and carry water from the roots to the leaves.

Detail Questions

1. A grapevine is a long, thin, curling, flexible stem.
2. A rosebush
3. Epidermis, cortex, and vascular cylinder
4. In a dicot plant it is a vascular cylinder; in a monocot plant it is many small, scattered tubes.

PRE-READING 5

Focus Question

The flower creates new plants; it is the reproductive system of the plant.

Detail Questions

1. Stamen, anther, pollen
2. Style, stigma, ovary, ovules
3. Pollen is carried by wind and insects.

CHAPTER | ZOOLOGY Mammals: Animals We Know Best

PRE-READING 1

Focus Questions

Plants can make their own food, but animals cannot.
Animals differ in what they eat, in their body temperature, and in their bone structure.

Detail Questions

1. An animal that eats grasses and other plants
2. An animal that eats other animals
3. Omnivores
4. Fish
5. 98.6° Fahrenheit or 37° Celsius
6. A vertebrate

PRE-READING 2

Focus Question

Some of the largest animals in size and the most familiar of the vertebrates are mammals.

Detail Questions

1. They have fur or hair and female mammals have milk-producing glands to feed their babies.
2. Their young are born from eggs. They are not born alive.
3. It is the center of thought in the brain.
4. The cerebrum in the human is larger than in other mammals.

PRE-READING 3

Focus Question

Mountains, deserts, water, grasslands

Detail Questions

1. Whales and dolphins
2. To get air to breathe
3. The male African elephant
4. The whale
5. They can store water in the humps in their backs.
6. Protective coloration, protective resemblance, mimicry

CHAPTER 3 | HUMAN ANATOMY Skin, Muscles, and Bones of the Human Body

PRE-READING 1

Focus Questions

Epidermis, dermis, hair follicles, sweat glands, oil glands
It keeps the body's water in, germs out, and helps the body feel things.

Detail Questions

1. Epidermis, dermis, and fatty tissue
2. A thick, hard place on the epidermis caused by rubbing over a long period of time
3. Pigment
4. Brown spots on the skin
5. Oil glands produce oil to keep skin and hair healthy; sweat glands help cool the body.
6. Keeping injured areas clean helps prevent infection.

PRE-READING 2

Focus Questions

Smooth muscles are found in the organs of the body.
Cardiac muscles are found in the heart.
Skeletal muscles are found all through the body, attached to bones.

Detail Questions

1. An organ is a part inside of the body that has a special function.
2. Examples of organs are the stomach, heart, liver, and lungs.
3. Biceps, for example, are voluntary muscles; the heart is an involuntary muscle.
4. Biceps, triceps
5. The biceps contracts and the triceps relaxes.

PRE-READING 3

Focus Questions

Cartilage, bone marrow, ligaments, tendons, joints
The skeletal system gives muscles a place to hold on to move the body, supports the body so it can stand up, and protects the organs of the body.

Detail Questions

1. The smallest bones are in the ears. The longest bone is in the upper leg.
2. Periosteum

3. Bone marrow makes new blood cells for the body.
4. Cartilage is a flexible material like bendable plastic; it is found in the ears and nose, for example.
5. Ligaments and tendons
6. Places where two bones are joined together

CHAPTER 4 | HUMAN PHYSIOLOGY   Digestion, Respiration, and Circulation

PRE-READING 1

Focus Question

Digestion is the breaking down, or changing, of food into a form that your body can use.

Detail Questions

1. Mechanical digestion is the breaking down of food by action of the teeth and the squeezing of the stomach muscles.
2. Chemical digestion is the breaking down of food by chemicals in the mouth, stomach, and intestines.
3. Absorption is the transfer of liquid food from the small intestines into the blood.
4. Saliva begins the process of chemical digestion.
5. Acids and enzymes
6. Food is absorbed into the blood.
7. Waste food is broken down and stored until it leaves the body.

PRE-READING 2

Focus Question

Respiration is the exchange of gases involved in breathing. Oxygen is breathed in and carbon dioxide is breathed out.

Detail Questions

1. Nose or mouth, pharynx, larynx
2. Larynx
3. To produce speech or sounds
4. Trachea
5. Lungs
6. Bronchi
7. Bronchioles
8. Oxygen, carbon dioxide
9. In the alveoli
10. A large muscle under the lungs that helps in breathing.

PRE-READING 3

Focus Question

Circulation is the pumping of blood around the body.

Detail Questions

1. Arteries, veins, capillaries
2. Arteries, veins
3. Plasma, white blood cells, red blood cells, platelets
4. White blood cells fight infection.

5. Blood carries food to your body, helps control body temperature, cleans waste from the body, protects against infection, takes in oxygen, and carries away carbon dioxide.

CHAPTER 5 — HUMAN PHYSIOLOGY The Sense Organs

PRE-READING 1

Focus Question

Vibrations, or sound waves, moving through the air, water, or ground

Detail Questions

1. Ear canal and eardrum
2. It vibrates.
3. Anvil, hammer, and stirrup
4. They look like these objects.
5. Moving liquid on the inside of the ear
6. The nerves pass the sound message on to your brain.

PRE-READING 2

Focus Question

The tongue and the nose

Detail Questions

1. Papillae
2. Taste buds or nerve cells that carry messages about the food you eat to the brain.
3. The front of the tongue
4. A small amount of gas
5. To bring an odor to the nose

PRE-READING 3

Focus Question

Iris, eyelid, pupil, and retina

Detail Questions

1. The iris is a small muscle that controls the amount of light that comes into your eye.
2. It changes when the light changes; it is large when the light is dim, small when the light is bright.
3. It controls the amount of light that comes in the eye and keeps the eye moist.
4. On the retina.
5. The brain receives the message from the eye and changes it to a picture or image.
6. Nearsightedness, farsightedness, or astigmatism
7. With eyeglasses or contact lenses

PRE-READING 4

Focus Question

They send messages to the brain.

Detail Questions

1. Pain, heat, cold, pressure, or light touch
2. Near hairs, hairless areas, or deep inside your body
3. Some parts of your body have more sensory nerves.
4. It helps you learn about your body, and it protects you.

CHAPTER 6 — HUMAN ECOLOGY   Healthful Living: Nutrition and Exercise

PRE-READING 1

Focus Question

Food with nutrients

Detail Questions

1. Promote growth, replace old cells, provide energy
2. Carbohydrates, fats, proteins, minerals, water, vitamins
3. Carbohydrates
4. Vitamins
5. Water
6. No single food has all the nutrients you need every day.

PRE-READING 2

Focus Question

To keep the body healthy

Detail Questions

1. Dairy group, protein group, fruit and vegetable group, bread and cereal group
2. Milk products
3. Cheese, ice cream, yogurt
4. They are a source of minerals and vitamins A and D.
5. Beans and peas
6. Protein group
7. Celery, grapes, apples, grains
8. It helps us to digest food.
9. Oats, rice, wheat, corn
10. Balanced diet

PRE-READING 3

Focus Question

Improves muscle tone, increases metabolic rate, controls body weight

Detail Questions

1. Strength, firmness, and good reaction time or response of muscles
2. Metabolic rate
3. Aerobic exercise is continuous, keeping the heartbeat fast and steady.
4. Weight lifting, walking slowly
5. At least 15 minutes, three times a week
6. Do the exercise you like best.

TO THE STUDENT

Life Science is the study of living things. You will study five different topics in this short introduction to life science.

Botany is the study of plants.

Zoology is the study of animals.

Human anatomy is the study of the structure of the human body.

Human physiology is the study of the organs, tissues, and cells that make up the human body.

Human ecology is the study of human beings in their world or environment.

> With this book you will learn to follow the scientific thinking process. You will do the following things:
>
> | | |
> |---|---|
> | **Consult with others:** | Share ideas in cooperative groups |
> | **Make hypotheses:** | Guess possible answers to questions or problems |
> | **Experiment:** | Try to prove your guesses were correct |
> | **Observe:** | Watch and take notes on what you see |
> | **Read:** | Learn new information and remember it |
> | **Classify:** | Put things or ideas into groups or categories |
> | **Compare and contrast:** | Discover how something is different from another or the same as another thing |
> | **Make conclusions:** | Decide if your guesses were correct or if you need to change your ideas |
> | **Report:** | Discuss your conclusions with the whole class |

You will also have the chance to ask and answer questions in the way scientists do. Life scientists ask these kinds of questions:

★ What are living things made of?
★ How are living things organized?
★ How do living things grow, change, and behave?
★ What do living things need to stay alive?
★ How do living things reproduce?
★ How do living things live in their world?

Of course, there are many more areas of life science to explore and discover. This book will give you a fine beginning in the study of living things.

Teacher's Edition

Life Science

Content and Learning Strategies

CHAPTER 1

BOTANY
ANGIOSPERMS: THE LARGEST CLASS OF PLANTS

INTRODUCTION

Botany is the study of plant life. Plants can be divided into groups or **classes**. In this chapter, you will learn about one class of plants called **angiosperms**. You will read about the different kinds of angiosperms, what angiosperms look like, how they make food, and what the functions of each part of the plant are.

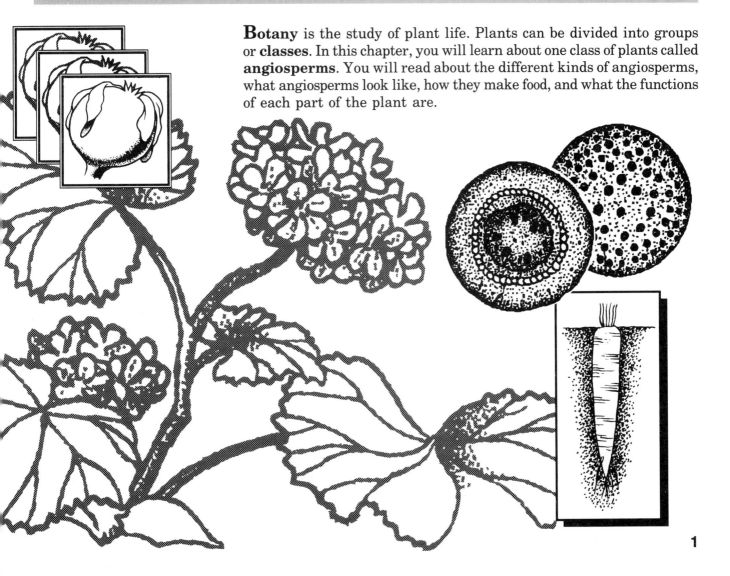

CRITICAL THINKING ACTIVITIES

WHAT DO YOU ALREADY KNOW ABOUT PLANTS?

LEARNING STRATEGIES
☆ Using prior knowledge
☆ Working cooperatively

Read these sentences. Draw a circle around the words you do not understand. Underline the words you cannot pronounce.

Plants grow in soil.
Soil has food and water that plants need.
All plants need water to live and grow.
Green plants need light to live and grow.
Some plants grow from seeds.

Sit down with a partner. Look at your book and your partner's book. Help each other understand the words that are circled. Help each other pronounce the words that are underlined.

THINK ABOUT THESE IDEAS

LEARNING STRATEGIES
☆ Imagery
☆ Taking notes
☆ Self-evaluation

Work in groups of three or four. Work together to answer these questions. If you are not sure about your answers, guess!

1. Look at the picture. Read the five words. Write the correct words on the lines.

   Flower   Stem   Leaf   Roots   Soil

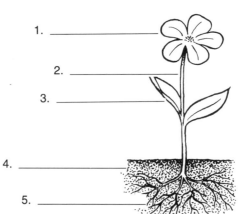

   1. _____
   2. _____
   3. _____
   4. _____
   5. _____

2. The plants you will be reading about in this chapter all have roots, stems, leaves, and flowers. Discuss with your partners the functions, or important jobs, of these plant parts. Write down your ideas in a notebook.

   a. Why are roots important to a plant? What is their important function?
   b. Why are leaves important to a plant? What is their important function?
   c. Why are stems important to a plant? What is their important function?
   d. Why are flowers important to a plant? What is their important function?

When your group finishes talking about these ideas, share your ideas with the whole class. Are your ideas different? Are they similar? After you read this chapter, look at these ideas and your answers again. Do not worry if your answers are right or wrong.

GROUP OBSERVATIONS

LEARNING STRATEGIES
☆ Imagery
☆ Working cooperatively
☆ Self-evaluation

Materials

small plants	tagboard or cardboard
a pair of scissors	a pen
tape	a ruler

Sit down in groups of three or four. In your group, cut the plants apart carefully. Separate the different parts. Put all the leaves together. Carefully tape them to the tagboard or cardboard. Write the word *leaves* in large letters. Do the same with the roots, the stem, and the flowers. Tape each of them to the tagboard. Write the names under each example. When you finish, share your results with the other groups. Then in your own group, compare the results of all the groups. Answer these questions. Remember, this is not a test. Guess if you are not sure.

1. Do all the roots look the same? Yes or no?
   How are they different?
   What colors are they?
   How long are they?
   How many roots are there for each plant?

2. Look at the stems. Are they the same? Yes or no?
   How are they different?
   What colors are they?
   How long are they?
   How thick are they?

3. Look at the leaves. Do they look the same? Yes or no?
   How are they different?
   What color are they?
   How big are they?

4. Look at the flowers. Do all these plants have flowers? Yes or no?
   How are they different?
   What color are they?
   How big are they?
   How many petals does the flower have?

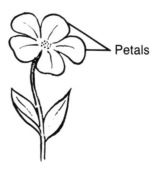

Share your answers with the whole class. After you finish reading this chapter, come back to these questions and observations and read them again. Are your answers the same?

PRE-READING 1

FOCUS QUESTION

Skim the reading on pages 4, 5, and 6 to find the answer to the question below. Underline the answer in your book. Write the answer below.

- ***What are angiosperms?*** _____
  _____

DETAIL QUESTIONS

LEARNING STRATEGY
☆ Reading selectively

Read "Characteristics of Angiosperms" on pages 4, 5, and 6. Find the details. Underline the answers in your book. Write the answers below. As you read, write down any words you do not understand or cannot pronounce on small slips of paper your teacher will give you. Then give these "vocabulary tickets" to your teacher. Do not write your name on the tickets. Your teacher and the whole class will review the words together.

1. Name the parts of an angiosperm. _____
   _____

2. What does the transport system do? _____
   _____

3. What are the two subclasses of angiosperms? _____
   _____

4. Name four differences between monocot and dicot plants.
   _____
   _____

READING 1 ★

Characteristics of Angiosperms

There are many **classes** of plants. Classes are like families or groups. This chapter is about the class of plants called **angiosperms**. Angiosperms are green plants that have flowers and produce seeds.

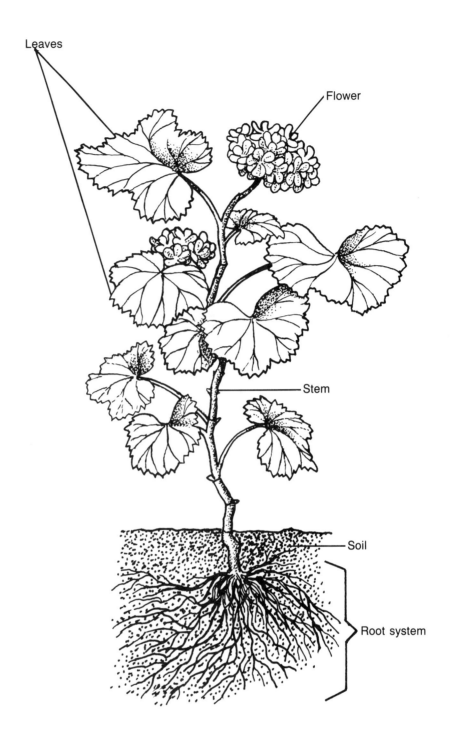

All angiosperms have the same parts: a root system, a stem, leaves, and flowers. These parts are all connected by a system of tiny tubes. This system of tiny tubes is called the **transport system**. The tubes carry water and food through the plant. The transport system is inside the stem, the roots, and the leaves.

Angiosperms are divided into two groups, or subclasses. One subclass is called **monocot** and the other is **dicot**. There are four important differences between monocots and dicots. You can see these differences in the picture that follows.

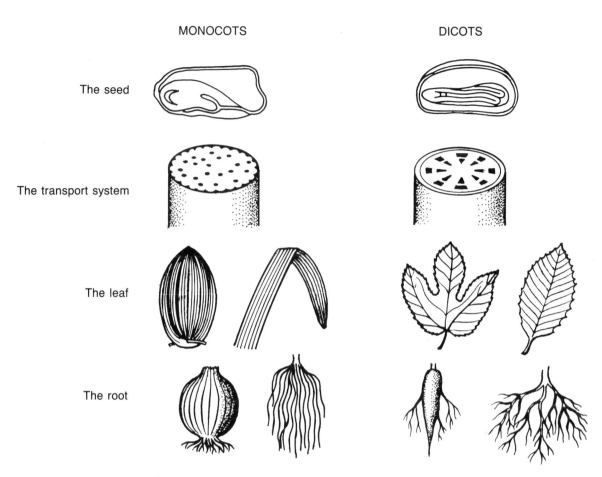

There are four differences between monocot and dicot plants.

The seeds in monocot and dicot plants are different. The seed for the dicot has two leaves. The seed for the monocot has only one leaf. The transport systems are also different. In the dicot plant, it is arranged around the center. In the monocot plant, it is scattered in small groups. The **veins** and roots are also different. The leaf vein looks like fans or feathers in dicots and parallel lines in monocots. Veins are small tubes in the leaves that carry the food and water. Roots in dicots are usually fat. Monocot roots are long, thin strings.

You will learn more about angiosperms and the differences between monocots and dicots in the rest of this chapter. You will read about the leaves, roots, stems, and flowers of angiosperms.

SELF-EVALUATION 1

VOCABULARY TICKETS Read the vocabulary tickets with your teacher and the whole class. Are there still some words you do not understand? Write these words in a notebook. With a partner, write some example sentences using these new words. Talk about the meaning of these words with your classmates.

VOCABULARY CHECK

Here are some important words from this reading. Do you understand all of these words? Circle the words you do not understand. Then find the words in the reading. Talk about the meaning of these words with your classmates. If you know all the words, continue to the Question Review.

angiosperms monocot
classes transport system
dicot veins

QUESTION REVIEW

Go back to the questions on page 4. Look at your answers. Work with a partner. Look at your partner's answers too. Are they the same as your answers? Help each other write the correct answers.

PRE-READING 2

FOCUS QUESTION

Skim the reading on pages 8 and 9 to find the answer to the question below. Underline the answer in your book. Write the answer below.

■ *What important process occurs within the leaf?* _____

_____

DETAIL QUESTIONS

LEARNING STRATEGY
☆ Reading selectively

Read "The Leaf" on pages 8 and 9. Find the details. Underline the answers in your book. Write the answers below. As you read, write down on your vocabulary tickets any words you do not understand or cannot pronounce.

1. What is a frond? _____

_____

2. Name two things that leaves collect. _____

_____

3. What gas does a leaf breathe in? _____

_____

4. In the plant, what process uses carbon dioxide, sunlight, and

water to make food? _____

_____

BOTANY ★ 7

5. What gas does a leaf breathe out? _____

6. What is the green material in a leaf called? _____

7. What does chlorophyll use to change water and carbon dioxide into sugar? _____

READING 2 ★

The Leaf

One part of a plant is the leaf. Leaves come in many different shapes and sizes. Some leaves are thick and broad. These are called **fronds**. Plants with fronds grow in rain forests where there is a lot of water, but very little sunlight. Other leaves are long and thin. Still others are small and hard. Plants with small, hard leaves grow in deserts where

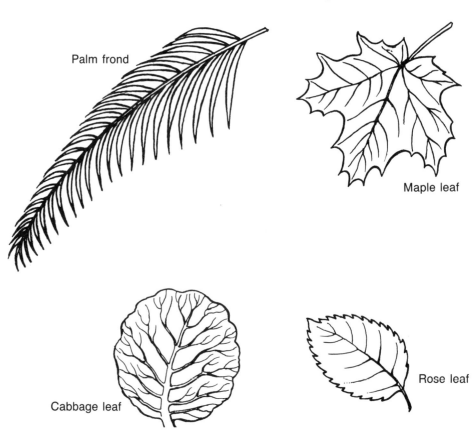

8 ★ CHAPTER 1

there is a lot of sunlight, but very little water. Whatever their shape and size, all leaves perform the same function for the plant.

All plants need sunlight to grow. The flat surface on the top of the leaf collects the sunlight for the plant. Plants also need **carbon dioxide**. The top surface of the leaf has **microscopic** pores, or tiny holes, that "breathe" or collect the carbon dioxide. The plant breathes in carbon dioxide and breathes out **oxygen**. Two functions of the leaf are to collect sunlight and carbon dioxide.

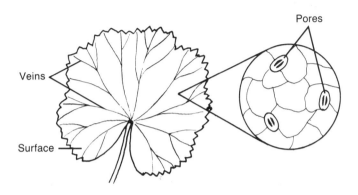

A very important process occurs within the leaf. It is called **photosynthesis**. Photosynthesis is a process in which plants use sunlight, carbon dioxide, and water to make food. Photosynthesis occurs in the green parts of plant leaves. This green material is called **chlorophyll**. Chlorophyll uses the energy from sunlight to change water and carbon dioxide into sugar. Plants need this sugar as food to grow.

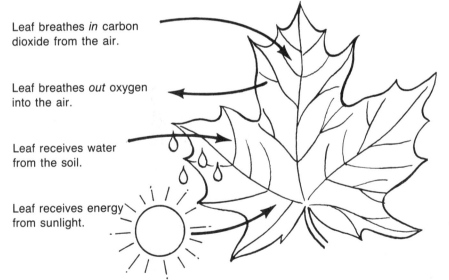

Photosynthesis. Chlorophyll in the leaf uses energy from the sun to change water and carbon dioxide into sugar for the plant.

In this reading you have learned that leaves come in different shapes and sizes, and perform important functions for the plant. In the rest of this chapter, you will learn about the other parts of the plant and their functions.

SELF-EVALUATION 2

VOCABULARY TICKETS

Read the vocabulary tickets with your teacher and the whole class. Are there still some words you do not understand? Write these words in a notebook. With a partner, write some example sentences using these new words. Talk about the meaning of these words with your classmates.

VOCABULARY CHECK

Here are some important words from this reading. Do you understand all of these words? Circle the words you do not understand. Then find the words in the reading. Talk about the meaning of these words with your classmates. If you know all the words, continue to the Question Review.

carbon dioxide microscopic
chlorophyll oxygen
fronds photosynthesis

QUESTION REVIEW

Go back to the questions on pages 7 and 8. Look at your answers. Work with a partner. Look at your partner's answers too. Are they the same as your answers? Help each other write the correct answers.

PRE-READING 3

FOCUS QUESTION

Skim the reading on pages 11 and 12 to find the answer to the question below. Underline the answer in your book. Write the answer below.

■ *Name two important functions of roots.* _____

_____

DETAIL QUESTIONS

LEARNING STRATEGY
☆ **Reading selectively**

Read "The Roots" on pages 11 and 12. Find the details. Underline the answers in your book. Write the answers below. As you read, write down on your vocabulary tickets any words you do not understand or cannot pronounce.

1. Name the three layers of a root. What is the function of each layer?

_____

_____

2. What does a taproot look like? _____

_____

3. What is the root in a dicot plant called? _____

_____

10 ★ CHAPTER 1

4. What do fibrous roots look like? _____

   _____

5. Does grass have fibrous roots or taproots? _____

   _____

READING 3 ★

The Roots

Roots perform functions for the plant. Roots pull water from the soil. They store, or keep, water and food until the plant needs them. They hold the plant up straight so it will not fall over. Roots also help keep the soil firmly packed around the plant so the soil cannot wash away in a rainstorm.

There are two different kinds of roots in plants. Dicot plants have **taproots**. A taproot is a large, fat root with smaller root hairs growing from it. Every time you eat a carrot or a radish you are eating a taproot. The roots in monocot plants are called **fibrous roots**. You can see these thin, stringlike roots when you pull up a handful of grass. If you go to the supermarket and look at a bunch of green onions, you will see fibrous roots.

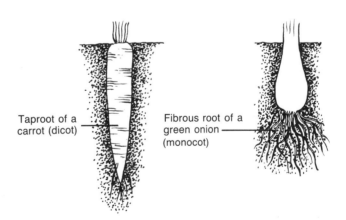

Types of root systems.

What does a root look like if you cut across it? The picture on the next page shows a cross section of a root. It shows the three layers that make up a root. You can easily see the three layers of a root if you cut a large carrot into slices.

1. The outside layer is like the skin. This skin is called the **epidermis**. The epidermis protects the root and helps it take in water.

BOTANY ★ 11

2. The **cortex** is the largest part inside the root. It stores water and food as it is made by the plant.

3. In the center of a root is a layer called the **vascular cylinder**. It is part of the transport system that moves the food and water up and around the plant.

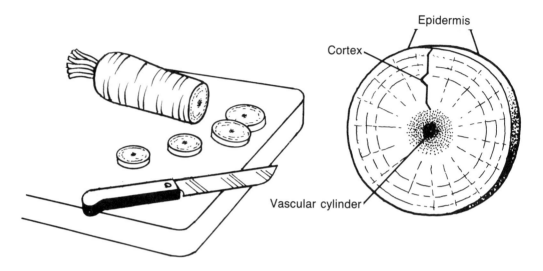

Roots have jobs they do for the plant. There are two types of roots and there are different layers in each root. In the next reading you will read about the stem of the plant and its important functions.

SELF-EVALUATION 3

VOCABULARY TICKETS Read the vocabulary tickets with your teacher and the whole class. Are there still some words you do not understand? Write these words in a notebook. With a partner, write some example sentences using these new words. Talk about the meaning of these words with your classmates.

VOCABULARY CHECK Here are some important words from this reading. Do you understand all of these words? Circle the words you do not understand. Then find the words in the reading. Talk about the meaning of these words with your classmates. If you know all the words, continue to the Question Review.

cortex taproots
epidermis vascular cylinder
fibrous roots

QUESTION REVIEW Go back to the questions on pages 10 and 11. Look at your answers. Work with a partner. Look at your partner's answers too. Are they the same as your answers? Help each other write the correct answers.

PRE-READING 4

FOCUS QUESTION

Skim the reading on pages 13, 14, and 15 to find the answer to the question below. Underline the answer in your book. Write the answer below.

■ *What are three important functions of stems?* _____

_____

DETAIL QUESTIONS

Read "The Stem" on pages 13, 14, and 15. Find the details. Underline the answers in your book. Write the answers below. As you read, write down on your vocabulary tickets any words you do not understand or cannot pronounce.

LEARNING STRATEGY
☆ **Reading selectively**

1. What does the stem of a grapevine look like? _____

_____

2. What kind of plant has a stem with thorns? _____

_____

3. What are the three layers, or parts, of a stem? _____

_____

4. What is the difference between the transport systems in the stems

of dicot and monocot plants? _____

_____

READING 4 ★

The Stem

There are many kinds of stems. The amaryllis plant has a tall, stiff stem. The stems of a grapevine are long and curling. These long and flexible stems are called **vines**. A grapevine also has **tendrils** on its stem. A tendril is a thin, curling "finger" that attaches the plant to a fencepost or a wire. Tendrils hold the vine and leaves up to the light. A pea plant is an example of a vine with tendrils.

Some stems run along the top of the soil, like the one on a strawberry plant. Other stems are very soft and thin, like the one on a tulip. And there are other stems that have **thorns**. A thorn is a sharp, pointed part of the stem. Rosebushes and berry bushes have sharp thorns on their stems.

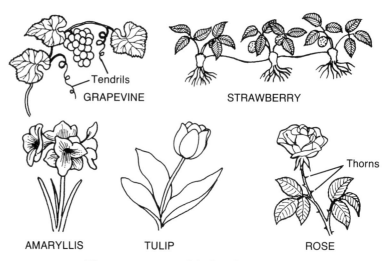

There are many kinds of stems.

Stems have three very important jobs, or functions. Stems hold the leaves up to the sunlight. Stems transport, or carry, the food from the leaves to the rest of the plant. Stems also take the water from the roots to the leaves.

What does a stem look like if you cut across it? A stem is made up of different layers similar to a root. Picture A shows the cross section of the stem of a dicot plant. Picture B shows the cross section of the stem of a monocot plant. You can easily see the layers that make up the stem in both plants. The outside layer is the epidermis. It is like a skin. The epidermis protects the stem. The cortex is the largest layer inside the stem. Stems also have a transport system. The transport system carries food and water to the leaves and the roots.

The transport systems in monocot and dicot plants are different. The transport system in dicot plants is made of a central tube (see picture A). It is called a vascular cylinder. In the monocot plant, the transport system is made up of many small tubes scattered in tiny groups throughout the cortex (see picture B).

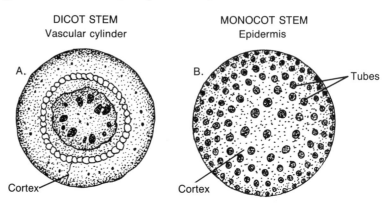

In this reading you learned about different kinds of stems and the important functions they perform. In the next reading you will learn about flowers and their special job.

SELF-EVALUATION 4

VOCABULARY TICKETS Read the vocabulary tickets with your teacher and the whole class. Are there still some words you do not understand? Write these words in a notebook. With a partner, write some example sentences using these new words. Talk about the meaning of these words with your classmates.

VOCABULARY CHECK Here are some important words from this reading. Do you understand all of these words? Circle the words you do not understand. Then find the words in the reading. Talk about the meaning of these words with your classmates. If you know all the words, continue to the Question Review.

    tendrils   vines
    thorns

QUESTION REVIEW Go back to the questions on page 13. Look at your answers. Work with a partner. Look at your partner's answers too. Are they the same as your answers? Help each other write the correct answers.

PRE-READING 5

FOCUS QUESTION Skim the reading on pages 16, 17, and 18 to find the answer to the question below. Underline the answer in your book. Write the answer below.

■ *What is the main function of a flower?* _____

_____

DETAIL QUESTIONS Read "The Flower" on pages 16, 17, and 18. Find the details. Underline the answers in your book. Write the answers below. As you read, write down on your vocabulary tickets any words you do not understand or cannot pronounce.

LEARNING STRATEGY
☆ Reading selectively

1. What are three parts of the male reproductive system of the flower?

_____

_____

BOTANY ★ 15

2. What are four parts of the female reproductive system of a flower?

   _____

   _____

3. Name two ways that pollen is carried to the stigma of a plant.

   _____

   _____

READING 5 ★

The Flower

The flower is the **reproductive system** of most seed plants. The flower makes it possible for the plant to make new plants or to **reproduce** itself. Most flowers have both male and female reproductive systems.

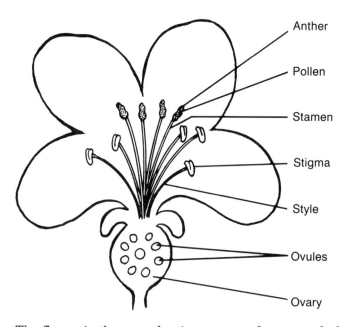

The flower is the reproductive system of most seed plants.

The **stamen, anther,** and **pollen** are the parts of the male reproductive system in the flower. Male reproductive cells are found in the dust, or powder, called pollen. Pollen is found in the anther. The anther is a small sac that holds the pollen. It is found at the end of the thin stalk called the stamen.

The parts of the female reproductive system of the flower are the **stigma, style, ovules,** and **ovary**. The female reproductive cells are in the ovules. The ovules are in the ovaries. These ovaries are attached to thin stalks called styles. At the end of the style is a stigma. The female stigma receives pollen from the male anther.

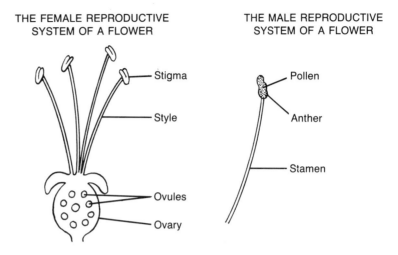

How do plants reproduce or make new plants? The pollen from the male anther falls on the female stigma. The pollen travels down the style to the ovules. When the pollen reaches the ovules, we say that the ovules have been **fertilized**. This process is called **pollination**. Many flowers pollinate themselves alone. Other flowers are **cross-pollinated**. Pollen from one flower is carried to another flower by the wind or by insects, such as bees and butterflies.

After the female cells are fertilized, or pollinated, a tiny plant called an **embryo** begins to develop inside each ovule. As the baby plant grows, the flower dries up.

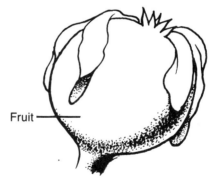

The outer parts of the flower dry up and fall off.
The ovary grows bigger, forms seeds, and becomes a fruit.

The ovary that is left begins to grow larger. This ovary is now called a **fruit**. We eat many fruits such as apples, peaches, plums, strawberries, oranges, and apricots. These fruits are **ripened** ovaries. Vegetables such as beans, cucumbers, and corn are also ripened ovaries. Nuts, like walnuts and almonds, are also the seeds from a ripened ovary.

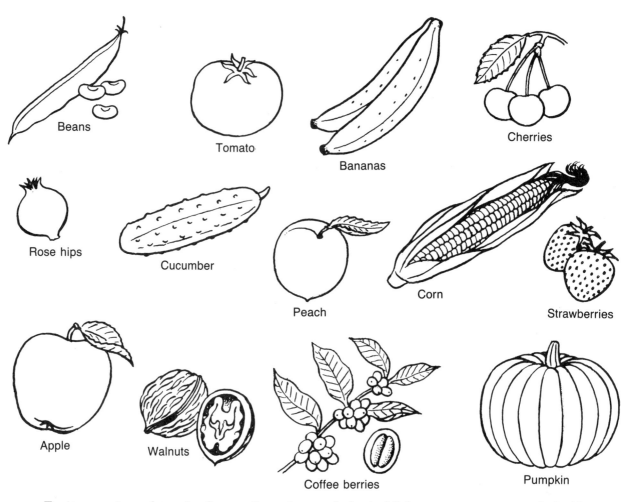

Fruits are ripened ovaries from a flowering seed plant. All have one or more seeds inside.

The flowers in monocot and dicot plants are different. The monocot flower usually has three stamens, six styles, and six **petals**. The dicot plant usually has flowers with four stamens, five styles, and five petals.

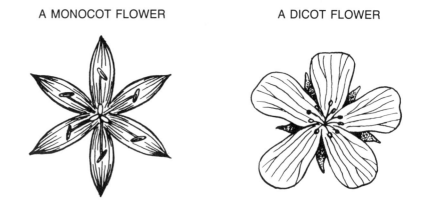

In this reading you learned about flowers and how they reproduce.

SELF-EVALUATION 5

VOCABULARY TICKETS Read the vocabulary tickets with your teacher and the whole class. Are there still some words you do not understand? Write these words in a notebook. With a partner, write some example sentences using these new words. Talk about the meaning of these words with your classmates.

VOCABULARY CHECK Here are some important words from this reading. Do you understand all of these words? Circle the words you do not understand. Then find the words in the reading. Talk about the meaning of these words with your classmates. If you know all the words, continue to the Question Review.

anther	ovules	reproductive system
cross-pollinated	petals	ripened
embryo	pollen	stamen
fertilized	pollination	stigma
fruit	reproduce	style
ovary		

QUESTION REVIEW Go back to the questions on pages 15 and 16. Look at your answers. Work with a partner. Look at your partner's answers too. Are they the same as your answers? Help each other write the correct answers.

CHAPTER REVIEW Now that you have completed your reading about plants, go back to pages 2 and 3. Look at your first ideas about plants. Have your ideas changed? What have you learned? Talk about your ideas with the teacher and the whole class.

EXTENSION ACTIVITIES

A. STUDENT ARTWORK Your teacher will give you a large piece of construction paper and some marking pens. Work with a partner. Draw three pictures.

LEARNING STRATEGY
☆ Imagery

1. Draw a cross section of a dicot stem.
2. Draw a cross section of a monocot stem.
3. Draw a cross section of a taproot.

For each picture, write the names of all the important parts and write what the job of each part is. Display your chart on the bulletin board.

B. SUPERMARKET

LEARNING STRATEGIES
☆ Grouping
☆ Taking notes

Go to the supermarket with a partner. Take a notebook and pencil. Look in the produce section. Write down the names of all the vegetables that are taproots, such as carrots and turnips. How many can you find? Next, write the names of all the vegetables that are leaves, such as lettuce and cabbage. If you are not sure of the names of the vegetables, read the signs or ask somebody in the store to help you.

C. FLOWER NURSERY

Go to a flower nursery with a partner. Look at the leaves of the plants. Write down the names of plants that are dicots. (Remember, leaves of dicot plants have veins that look like feathers or fans.) Write down the names of plants that are monocots. (Their leaves have veins that are in straight lines, side by side.)

D. MONOCOTS AND DICOTS

You and your classmates will each bring a leaf to class. Separate the monocot leaves from the dicot leaves. Observe the differences in the veins. Put several layers of newspaper on the table. Put the leaves on top of the newspaper so that the leaves do not touch. Cover the leaves with more newspaper. Put a heavy board or books on top of the newspaper. Keep the leaves in this "press" until they dry. When the leaves are dry, arrange them in a decorative way on tagboard or construction paper, or decorate a bulletin board.

E. SWEET POTATOES AND AVOCADOS

Materials

| a sweet potato or an avocado seed   toothpicks |
| a glass or jar of water |

Plant a sweet potato or an avocado seed in a glass of water. Only the bottom third of the seed or root should be in the water. Poke toothpicks into the seed or root and rest the toothpicks on the jar or glass rim to hold the plant out of the water. Roots will grow first. Later, the stem and leaves will form.

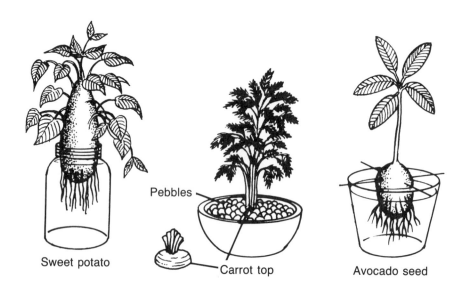

Sweet potato — Pebbles — Carrot top — Avocado seed

Roots of carrots, beets, and turnips can also be planted in water. Remove the old leaves from the tops of these roots. Cut off all the roots except one inch. Set these root pieces in shallow bowls of water. Put a few small rocks on top of the roots to hold them in place. New leaves will sprout from the tops of these roots.

F. SEED SPROUTING EXPERIMENT

LEARNING STRATEGY
☆ Taking notes

Materials

| a clear plastic cup | three or four lima bean seeds |
| paper towels | sand |

Concept:

A seed needs water to grow.

Work in groups of three or four. Make a tube from a piece of paper towel and fit it inside a plastic cup. Fill the cup with sand. Add water to moisten the sand. The paper towel should be moist, too. Place three or four seeds in the cup between the paper towel and the plastic. The wet sand against the sides of the plastic will hold the seeds in place. Observe the seeds each day. Keep the sand and the paper towel moist. Record your observations for two weeks. Be prepared to share your responses with the class.

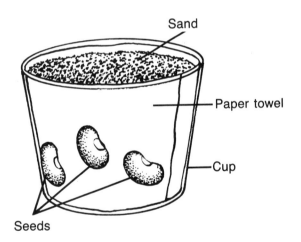

GLOSSARY

angiosperms Green plants that have flowers and produce seeds.

anther A small sac on the end of the flower's stamen.

botany The study of plant life.

carbon dioxide A gas in the atmosphere that plants breathe in.

chlorophyll The green material in the leaves of plants.

classes Special, distinct groups of living things.

cortex The largest inside part of a root or stem.

cross-pollinated A plant is cross-pollinated when pollen from one flower moves to another flower.

dicot A subclass of angiosperm plants.

embryo The baby plant inside a seed.

epidermis The outside protective covering of a root or stem.

fertilized A plant is fertilized when the male and female cells of the plant join together.

fibrous roots Thin, stringlike roots.

fronds Large, broad leaves.

fruit The ripened ovary of a flower.

microscopic Very small, not able to be seen with the eye alone.

monocot A subclass of angiosperm plants.

ovary A part of the female reproductive system in a plant.

ovules Female reproductive cells inside the ovary of a flower.

oxygen A colorless gas that plants breathe out into the atmosphere.

petals Parts of a flower.

photosynthesis A process by which plants use sunlight, carbon dioxide, and water to make food.

pollen A powdery material found in the anther of a flower; it contains the male reproductive cells.

pollination Fertilization; the joining of male and female reproductive cells in a plant.

reproduce To multiply or make new plants.

reproductive system The parts of a flower that are needed to make new plants.

ripened Became mature.

stamen The thin stalk of the male reproductive system of a flower.

stigma The top part of the style.

style The thin, tiny stalk inside a flower.

taproots Thick, fat roots.

tendrils Thin, curling parts on some vine stems.

thorns Sharp, pointed parts of some stems.

transport system A system of tubes that carries food and water through a plant.

vascular cylinder The part of the transport system that moves food and water up and around the plant.

veins Small tubes in the leaves that carry food and water for a plant.

vines Long, flexible stems.

CHAPTER 2

ZOOLOGY
MAMMALS: ANIMALS WE KNOW BEST

INTRODUCTION

Zoology is the study of animals. In this chapter you will read about the differences in animals. Then you will read about one class, or group, of animals called **mammals**. You will read about what they eat, what they look like, how their bodies work, where they live, and what makes them different from other animals.

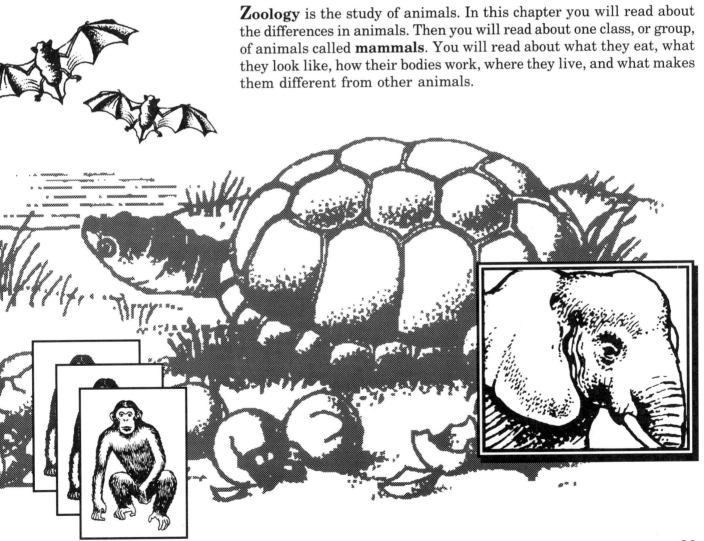

CRITICAL THINKING ACTIVITIES

WHAT DO YOU ALREADY KNOW ABOUT ANIMALS?

Read these sentences. Draw a circle around the words you do not understand. Underline the words you cannot pronounce.

LEARNING STRATEGIES
☆ Using prior knowledge
☆ Working cooperatively

Animals eat plants and other animals.
Animals cannot make their food inside their bodies.
Some animals have backbones.
Some of the largest animals are mammals.
Humans are mammals.

Sit down with a partner. Look at your book and your partner's book. Help each other understand the words that are circled. Help each other pronounce the words that are underlined.

THINK ABOUT THESE IDEAS

Work in groups of three or four. Work together to answer these questions. If you are not sure about your answers, guess!

LEARNING STRATEGIES
☆ Using prior knowledge
☆ Inferencing
☆ Self-evaluation

1. Look at the list of animals below. If you think the animal eats meat, write the word *meat* in the blank next to the animal. If you think the animal eats plants, write the word *plants* in the blank next to the animal.

    _____ cat   _____ horse

    _____ cow   _____ skunk

    _____ dog   _____ goat

    _____ sheep   _____ wolf

    _____ elephant   _____ lion

    _____ kangaroo   _____ tiger

2. Do all animals have backbones? Yes or no? If you answered no, write the name of an animal without a backbone.

    _____

3. The young of some animals are born alive. Other animals come out of eggs. Put a check (✓) beside the animals that are born alive.

    ____ bears   ____ cows   ____ snakes   ____ turtles

    ____ horses   ____ birds   ____ whales   ____ fish

4. Some animals live in the **desert**. Name an animal that lives in the desert. _____

5. Some animals live in the mountains. Name an animal that lives in the mountains. _____

6. Some animals live in the flat **grasslands**. Name an animal that lives in the flat grasslands. _____

When your group finishes talking about these ideas, share your ideas with the whole class. Are your ideas different? Are they similar? After you read this chapter, look at these ideas and your answers again. Do not worry if your answers are right or wrong.

GROUP OBSERVATIONS

LEARNING STRATEGIES
☆ Working cooperatively
☆ Self-evaluation

Sit down in groups of three or four. With your group make a list of ten different animals. Write the names of the animals here.

1. _____ 6. _____

2. _____ 7. _____

3. _____ 8. _____

4. _____ 9. _____

5. _____ 10. _____

From the list above, choose two animals.

1. _____ 2. _____

Write five differences between these two animals.

1. _____

2. _____

3. _____

4. _____

5. _____

ZOOLOGY ★

Share your answers with the whole class. After you finish reading this chapter, come back to these questions and observations and read them again. Are your answers the same?

PRE-READING 1

FOCUS QUESTIONS

Skim the reading on pages 27, 28, and 29 to find the answers to the questions below. Underline the answers in your book. Write the answers below.

- *What is the most important difference between plants and animals?* _____

- *What are three important differences among animals?*

DETAIL QUESTIONS

LEARNING STRATEGY
☆ Reading selectively

Read "Differences in Animals" on pages 27, 28, and 29. Find the details. Underline the answers in your book. Write the answers below. As you read, write down on your vocabulary tickets any words you do not understand or cannot pronounce.

1. What is a herbivore? _____

2. What is a carnivore? _____

3. Are humans omnivores or carnivores? _____

4. Name a cold-blooded animal. _____

5. What is a person's normal body temperature? _____

6. Is a horse a vertebrate or an invertebrate? _____

READING 1 ★★★★★★★★★★★★★★★★★★★★★★★★★★

Differences in Animals

Plants and animals are both living things. Animals are different from plants because animals can move around. There is a much more important difference. In Chapter 1, you learned that plants make their own food. Animals cannot make their food inside their bodies. Animals must eat plants or other animals for their food. Some animals eat only plants. Animals that eat only plants are **herbivores**. Cows are herbivores

Which animals are carnivores? Which animals are herbivores?

because they eat different kinds of grasses and grains. They do not eat other animals. Some animals eat the meat of other animals. These animals are **carnivores**. Lions, tigers, bobcats, and hawks are carnivores because they eat other animals like deer and mice. Some animals eat both plants and other animals. They are **omnivores**. Humans are omnivores because they can eat meat, fruits, grains, and vegetables.

There are other important differences among animals. Another difference among animals is their **body temperature**. Some animals are **cold-blooded**. A cold-blooded animal cannot control its body temperature. The body temperature of a cold-blooded animal changes with the temperature of the air or water around it. Fish are cold-blooded animals. The body temperature of a fish changes. Fish can sometimes stay frozen in a lake in cold weather. When the ice in the lake melts, the fish swims away.

Other animals are **warm-blooded**. Warm-blooded animals have a constant body temperature. The body temperature of a warm-blooded animal will stay the same even when the temperature in the air or water around it changes. Humans are warm-blooded. A person's normal body temperature is 98.6° **Fahrenheit** or 37° **Celsius**. The body temperature of a person may change if he or she is sick or falls into an icy lake. Warm-blooded animals must have protection to keep their body temperature from changing. Humans wear warm clothing in the cold winter months to protect their bodies from the cold. Animals like cats and dogs have thick skin and fur to protect them from the cold. Some animals grow thicker fur when it is cold.

Some animals are cold-blooded like fish and snakes. Other animals are warm-blooded like dogs and cats.

Another important difference among animals is the **spinal column** or **backbone**. Some animals like clams, butterflies, and bees do not have spinal columns. They are **invertebrates**. (From the Latin *in* meaning *not* and *vertebrate* meaning *bone*.) **Vertebrates** are animals that have spinal columns, for example, dogs, horses, elephants, and monkeys.

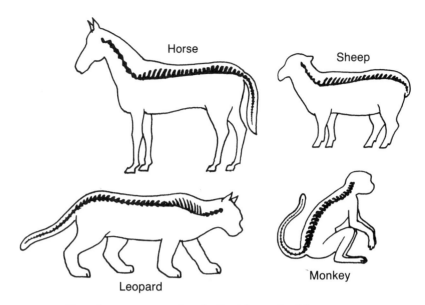

Vertebrates are animals that have backbones.

In this reading you learned that there are differences among animals. Animals eat different types of food, they have different body temperatures, and some animals do not have backbones. In the next reading you will learn about mammals and how they are different from other animals.

SELF-EVALUATION 1

VOCABULARY TICKETS Read the vocabulary tickets with your teacher and the whole class. Are there still some words you do not understand? Write these words in a notebook. With a partner, write some example sentences using these new words. Talk about the meaning of these words with your classmates.

VOCABULARY CHECK Here are some important words from this reading. Do you understand all of these words? Circle the words you do not understand. Then find the words in the reading. Talk about the meaning of these words with your classmates. If you know all the words, continue to the Question Review.

backbone	herbivores
body temperature	invertebrates
carnivores	omnivores
Celsius	spinal column
cold-blooded	vertebrates
Fahrenheit	warm-blooded

QUESTION REVIEW Go back to the questions on page 26. Look at your answers. Work with a partner. Look at your partner's answers too. Are they the same as your answers? Help each other write the correct answers.

PRE-READING 2

FOCUS QUESTION

Skim the reading on pages 30, 31, and 32 to find the answer to the question below. Underline the answer in your book. Write the answer below.

■ *What group of vertebrates do some of the largest animals belong to?* _____

_____

DETAIL QUESTIONS

LEARNING STRATEGY
☆ Reading selectively

Read "Characteristics of Mammals" on pages 30, 31, and 32. Find the details. Underline the answers in your book. Write the answers below. As you read, write down on your vocabulary tickets any words you do not understand or cannot pronounce.

1. What are the two main characteristics that make mammals different from other animals? _____

_____

2. How are the duck-billed platypus and the spiny anteater different from other mammals? _____

_____

3. What is the cerebrum? _____

_____

4. Why are humans considered the most intelligent of mammals?

_____

_____

READING 2 ★

Characteristics of Mammals

Some of the largest animals belong to the class of vertebrates called **mammals**. This group of animals includes many familiar animals such as dogs, cats, whales, elephants, monkeys, and even humans. Mammals have two main characteristics that make them different from other animals. First, they all have fur or hair. Second, female mammals make milk to feed their babies. The milk is made in the mother's body in glands called **mammary glands**. All newborn mammals are completely dependent on their mothers. Mother's milk is needed for the newborn mammals to survive.

This baby bison is drinking milk from its mother.

Most mammals are born alive. Only two mammals are not born alive. They are the duck-billed platypus and the spiny anteater. These two mammals lay eggs. Their babies are not born alive. When the eggs **hatch**, or break open, the babies get milk from the mother.

Baby turtles are born in eggs. They are not born alive. When the eggs hatch, the babies do not get milk from the mother turtle. Turtles are not mammals.

Mammals are the most intelligent, or smartest, of the vertebrates and humans are considered the most intelligent of mammals. Humans are considered the most intelligent because the human **cerebrum** is one of the largest of all mammals in comparison to the rest of the brain. The cerebrum is the part of the brain that is the center of thought. The cerebrum in the brain of mammals is larger than that of other kinds of animals. This is why mammals like dogs and cats are considered more intelligent than birds and snakes.

One measure of intelligence is the ability to learn new things. Humans learn new things quickly because they are intelligent. Other mammals that learn new things quickly are whales, dolphins, chimpanzees, and apes. This means that these mammals are also intelligent.

There are many different groups of mammals. They come in all sizes and shapes. Almost all mammals belong to the twelve orders of mammals.

Twelve orders of mammals

32 ★ CHAPTER 2

In this reading you learned about mammals and how they are different from other animals. You also learned that mammals all have hair or fur and female mammals have milk-producing glands. In the next reading, you will continue reading about mammals. You will learn where mammals live.

SELF-EVALUATION 2

VOCABULARY TICKETS Read the vocabulary tickets with your teacher and the whole class. Are there still some words you do not understand? Write these words in a notebook. With a partner, write some example sentences using these new words. Talk about the meaning of these words with your classmates.

VOCABULARY CHECK Here are some important words from this reading. Do you understand all of these words? Circle the words you do not understand. Then find the words in the reading. Talk about the meaning of these words with your classmates. If you know all the words, continue to the Question Review.

cerebrum   mammals
hatch   mammary glands

QUESTION REVIEW Go back to the questions on page 30. Look at your answers. Work with a partner. Look at your partner's answers too. Are they the same as your answers? Help each other write the correct answers.

PRE-READING 3

FOCUS QUESTION Skim the reading on pages 34, 35, and 36 to find the answer to the question below. Underline the answer in your book. Write the answer below.

■ *Name four habitats where we find mammals.* _____

_____

DETAIL QUESTIONS Read "Habitats of Mammals" on pages 34, 35, and 36. Find the details. Underline the answers in your book. Write the answers below. As you read, write down on your vocabulary tickets any words you do not understand or cannot pronounce.

LEARNING STRATEGY
☆ Reading selectively

1. Name two water mammals. _____

_____

2. Why do whales come to the surface of the water? _____

_____

ZOOLOGY ★ 33

3. What is the largest land mammal? _____

_____

4. What is the largest water mammal? _____

_____

5. Why are camels used for travel in the desert? _____

_____

6. Name three ways animals adapt to their habitats. _____

_____

READING 3 ★

Habitats of Mammals

Mammals live in many different places, or **habitats**. A habitat is the place an animal is usually found. Most mammals live on land, but some mammals live in the water. Whales and dolphins live in the water. They look like fish, but they are not fish. Whales and dolphins breathe air. They can only go into deep water for a short time. Then they have to come up to the surface for air. The babies of water mammals are born alive, and the females make milk to feed them. Whales are the largest mammals.

Whales and dolphins are water mammals that look like fish.

Some mammals like sheep, mountain goats, and leopards live in the mountains. These animals are good climbers. Animals who live in the mountains must be able to climb up and down the mountains to find food. Many mountain animals also have thick, fur coats to keep them warm in the cold, mountain weather.

Giraffes, zebras, elephants, and deer are all found in the flat **grasslands**. Most of these animals eat grass and leaves. The largest of all land mammals is the elephant. A male African elephant can weigh up to six tons, or 12,000 pounds. (A **ton** is the unit of measure for 2,000 pounds.) It has a long **trunk** that it uses to pick up objects, to spray mud on its back for protection from insects and the hot sun, and to carry water to its mouth.

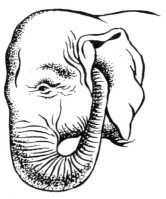
This is how the elephant cleans its ears with its trunk.

This elephant is spraying dust on its back with its trunk.

This is how the elephant drinks water from its trunk.

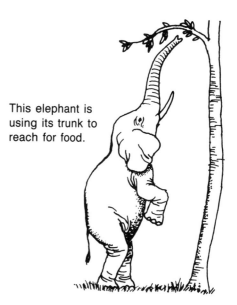
This elephant is using its trunk to reach for food.

The elephant uses its trunk for many things.

Other mammals live in the **desert**. The mammals that live in the desert must be able to live for many days with very little food and water. People use camels to travel through the desert. Camels can go for many days without drinking water. The desert camel can do this because it stores water in its **humps**.

Camels store water in the humps on their backs. Some camels have only one large hump; others have two smaller humps.

Mammals can be found in all parts of the world. They can be found anywhere from the hot, dry deserts to the cold, windy mountain ranges. They must adapt, or change, in order to live safely and protect themselves in their habitats.

One way mammals protect themselves is with their color. This is called **protective coloration**. Some animals are colored to look like their background. For example, a polar bear has a white coat. It is difficult to see a polar bear in the snow. A baby deer has a brown, spotted coat. It is difficult to see a baby deer in the forest. Mammals use protective coloration to protect themselves. In addition to protective coloration, other animals use **protective resemblance** and **mimicry** to protect themselves. For example, the stick insect looks just like a twig or small stick. This is called protective resemblance. Animals that eat the stick insect may think it is a stick and not eat it.

Sometimes one animal looks like another animal. This is called mimicry. For example, wasps are black and yellow. They are dangerous and unpleasant to eat. Hoverflies are also black and yellow, but they are harmless and edible. Birds often think the hoverflies are wasps and leave them alone.

In this reading you learned that mammals are found in different habitats such as water, mountains, flat grasslands, and desert. Animals adapt to their habitats in different ways.

SELF-EVALUATION 3

VOCABULARY TICKETS Read the vocabulary tickets with your teacher and the whole class. Are there still some words you do not understand? Write these words in a notebook. With a partner, write some example sentences using these new words. Talk about the meaning of these words with your classmates.

VOCABULARY CHECK Here are some important words from this reading. Do you understand all of these words? Circle the words you do not understand. Then find the words in the reading. Talk about the meaning of these words with your classmates. If you know all the words, continue to the Question Review.

desert	humps	protective resemblance
grasslands	mimicry	ton
habitats	protective coloration	trunk

QUESTION REVIEW Go back to the questions on pages 33 and 34. Look at your answers. Work with a partner. Look at your partner's answers too. Are they the same as your answers? Help each other write the correct answers.

CHAPTER REVIEW Now that you have completed your reading about mammals, go back to pages 24, 25, and 26. Look at your first ideas about mammals. Have your ideas changed? What have you learned? Talk about your ideas with the teacher and the whole class.

EXTENSION ACTIVITIES

A. SEARCH AND FIND

LEARNING STRATEGY
☆ Using resources

Find five pictures of animals in magazines. Glue each one to a piece of paper. With your partners, write a short summary about each animal. Use the following questions to help you. If you do not know the answers to the questions, find the information by asking your classmates or your teacher. You can also look up the information in a dictionary or encyclopedia.

Questions

1. What is the name of the animal?
2. Is it a vertebrate or an invertebrate?
3. Is it a herbivore, omnivore, or carnivore?
4. What things does it eat?
5. What is its habitat?
6. Is it a mammal? If yes, what order does it belong to?
7. Give other information that you know or can find out about this animal.

B. ANIMAL CATEGORIES

LEARNING STRATEGIES
☆ Grouping
☆ Working cooperatively
☆ Using resources

Sit down in groups of three or four. Use only *one* paper and *one* pencil. Put the animals from the list below into the categories that follow. Animals can fit into more than one category. If no one in your group knows the animal, ask your teacher or look it up in the dictionary or encyclopedia.

whale	dog	cat
elephant	lion	ape
tiger	bear	raccoon
monkey	seal	giraffe
wolf	mountain lion	anteater
zebra	deer	mouse
cow	buffalo	mountain goat
camel	bat	duck-billed platypus
armadillo	sloth	rat
pig	sheep	gerbil
dolphin	kangaroo	

Categories

water mammals:
grassland mammals:
mountain mammals:
desert mammals:
carnivores:
herbivores:
egg-laying mammals:
omnivores:

Now, think of five more animals that will fit into these categories. Share your list with the rest of the class. Do your classmates know the animals you know?

GLOSSARY

backbone Another word for spinal column.

body temperature A measure of warmth or coldness in the body.

carnivores Animals that eat other animals.

Celsius A temperature scale where 0° is freezing and 100° is boiling.

cerebrum The center for thought in the brain.

cold-blooded Animals that cannot regulate their body temperature are called cold-blooded animals.

desert An area of land with very little water.

Fahrenheit A temperature scale where 32° is freezing and 212° is boiling.

grasslands Land that is flat and wide and covered by grasses.

habitats The distinct places where different animals live.

hatch To break open; to come out of an egg.

herbivores Animals that eat plants.

humps Rounded bumps on the back of a camel that are used for storing water.

invertebrates Animals that don't have a backbone.

mammals A class of animals with hair or fur and mammary glands. Animals that feed milk to their babies.

mammary glands The parts of an animal's body that can produce milk to feed its babies.

mimicry Looking like something else; animals that look like other animals.

omnivores Animals that eat both animals and plants.

protective coloration When an animal has a color that makes it look like a part of its habitat (like a tree or a rock), it has protective coloration.

protective resemblance When an animal has a shape that makes it look like a part of its habitat (such as a flower or a rock), it has protective resemblance.

spinal column The backbone; a series of bones running down the back of an animal.

ton A unit of measure for 2,000 pounds.

trunk The long nose on an elephant.

vertebrates Animals that have a spinal column.

warm-blooded Animals that can regulate their body temperature are called warm-blooded animals.

zoology The study of animals.

CHAPTER 3
HUMAN ANATOMY
SKIN, MUSCLES, AND BONES OF THE HUMAN BODY

INTRODUCTION

Human anatomy is the study of the structure of the human body. This chapter will include readings about three main parts of the body: skin, **muscles**, and the **skeletal system** (bones). Skin covers your whole body and helps the body in important ways. The muscles and bones in the body make up over half of the body's weight. Muscles help the body move and make up a part of many important organs inside the body. The skeletal system is made up of many parts, and it works with the muscles.

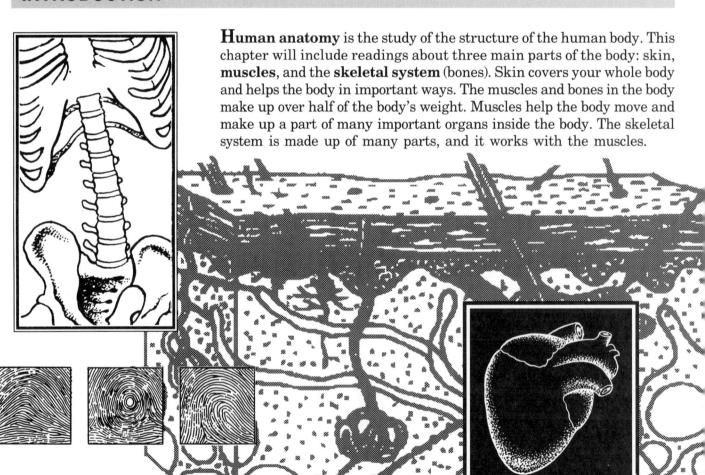

CRITICAL THINKING ACTIVITIES

WHAT DO YOU ALREADY KNOW ABOUT SKIN, MUSCLES, AND BONES?

LEARNING STRATEGIES
☆ Using prior knowledge
☆ Working cooperatively

Read these sentences. Draw a circle around the words you do not understand. Underline the words you cannot pronounce.

People's skin can be different colors.
Most skin is covered by hair.
Skin has glands that release sweat.
Bones hold up our bodies.
Some bones are large and others are small.
There are different kinds of muscles.

Sit down with a partner. Look at your book and your partner's book. Help each other understand the words that are circled. Help each other pronounce the words that are underlined.

THINK ABOUT THESE IDEAS

LEARNING STRATEGIES
☆ Inferencing
☆ Self-evaluation

Work in groups of three or four. Work together to answer these questions. If you are not sure about your answers, guess!

1. Why do you need skin? What does skin do for the body?
2. Why do you need bones? What do bones do for the body?
3. Why do you need muscles? What are the functions of muscles?
4. Does brushing your hair make it shiny? Why or why not?
5. Can bones bend? Yes or no? If you answered yes, where do you think these bendable bones in the body are?
6. Where are the smallest bones, and where are the longest bones in your body?
7. Children often fall down when they play. Why do their bones seldom break?

When your group finishes talking about these ideas, share your ideas with the whole class. Are your ideas different? Are they similar? After you read this chapter, look at these ideas and your answers again. Do not worry if your answers are right or wrong.

GROUP OBSERVATIONS

LEARNING STRATEGIES
☆ Inferencing
☆ Taking notes
☆ Self-evaluation

Sit down in groups of three or four. Read the questions together and talk about what you see. Write down some ideas. This is not a test. Write anything you see, guess, or imagine.

1. Your group will need a potato and a potato peeler. Peel the skin off half the potato. Write down how the peeled half looks and feels. Write down how the unpeeled half looks and feels. Make notes about how the potato will look tomorrow.

   The unpeeled half looks _____.

   It feels _____.

The peeled half looks _____.

It feels _____.

Tomorrow, the peeled half of the potato will look _____
_____ and feel _____
_____.

2. Your group will need a magnifying glass. Pull a few long hairs out of your head. Find one with a soft, white ball on the end. Look at this white ball under the magnifying glass.

   What do you see? _____
   _____

   What do you think the white ball is? _____
   _____

   What is it for? _____
   _____

3. Find the place where two bones come together in your fingers. These are called **joints**. Find the joint in your elbow, knee, wrist, and ankle. Touch them and move them. Write down some notes in your notebook about what you see and feel in your joints.

4. Touch your ears. Move them and bend them. Pinch them and twist them gently. Do the same with your nose. What do you feel in these two places on your body? Draw a circle around the following words that describe how your nose and ears feel when you bend and move them around.

hard	soft	straight	bendable	like plastic
like wood	stiff	weak	round	curved
breakable	strong	like paper	smooth	bumpy

5. Look at the hand you write with. Look at the place on your finger where your pencil presses when you write. What do you see there? Is everyone's finger the same? Why does your finger look like that?

6. Look on your hands and arms for brown spots, either **freckles** or **moles**. Look at them under the magnifying glass. What do you see? What causes these brown spots?

7. You will need to look at a clock or watch for this "experiment." Make a fist with your little finger curled up. Then move your little finger straight out. Do this two or three times for practice. Then, looking at the clock, count how many times each person in your

group can do this in *one minute*. Take turns. Write down your group members' names. Write down their scores next to their names.

> Was this an easy movement for you? Yes or no?
> Did your finger get tired? Yes or no?
> Were some students able to do it many more times than other students? Yes or no? Why?

When your group finishes making these observations and talking about the questions, share your observations with the whole class. Are your observations different? After you finish reading this chapter, come back to these questions and observations and read them again. Are your answers the same?

PRE-READING 1

FOCUS QUESTIONS

Skim the reading on pages 44, 45, and 46 to find the answers to the questions below. Underline the answers in your book. Write the answers below.

■ *Name five parts of the skin.* _____

_____

■ *How does skin protect the body?* _____

_____

DETAIL QUESTIONS

Read "The Skin" on pages 44, 45, and 46. Find the details. Underline the answers in your book. Write the answers below. As you read, write down on your vocabulary tickets any words you do not understand or cannot pronounce.

LEARNING STRATEGY
☆ Reading selectively

1. What are the three layers of skin called? _____

   _____

2. What is a callus? _____

   _____

3. What causes different people to have skin of different colors?

   _____

   _____

4. What are freckles and moles? _____

   _____

HUMAN ANATOMY ★ 43

5. What two types of glands are found in the skin? What are their functions? _____

_____

6. Why do we need to keep cuts or scratches clean? _____

_____

READING 1 ★ ★ ★ ★ ★ ★ ★ ★ ★ ★ ★ ★ ★ ★ ★ ★ ★ ★

The Skin

The human body is covered with skin. Human skin is **elastic**; it can stretch, move, and bend. The skin is made up of different parts. Skin has three layers. The outside layer is called the **epidermis**. The surface of the epidermis is not alive. This means that there is no blood in the epidermis. It is dry, dead skin. The epidermis protects the other layers of skin.

If the epidermis is constantly rubbed in one place, it will become thicker and harder. These thick, hard places are called **calluses**. You may have calluses on your hands, on the bottoms of your feet, or on the backs of your heels. Gardeners and carpenters have many calluses on their hands from using shovels and hammers. Waitresses and mailcarriers have calluses on their feet from walking every day. You may have a callus on the middle finger of your writing hand from writing every day.

The epidermis contains **pigment**, the material that gives skin its color. The color of your skin depends on how much pigment you have in it. All people have brown and yellow pigment in their skin. Everybody has a different amount of pigment. Black skin has more brown pigment and very little yellow pigment. Yellow skin has more yellow pigment and very little brown pigment. White skin has very little brown or yellow pigment. The amount of pigment you have in your skin depends on the amount your parents have.

Many people are born with some colored areas on their skin called **freckles** and **moles**. Freckles and moles are caused by the presence of pigment in one place on your skin. Some freckles can be caused by sunlight.

The next layer of skin is called the **dermis**. The dermis is made of **connective tissue**, a thin, flat tissue that holds the other tissues together. There are small tubes called **blood vessels** in the dermis. They carry the blood. There are also **nerve endings** in the dermis. Nerve endings help you to feel hot and cold on your skin.

The third layer of skin is made of fatty tissue. It is this layer that protects the body from bumps. Skin is made up of many parts. The epidermis, pigment, dermis, and blood vessels are only four parts of your skin. Look at the picture. You will see other parts of the skin.

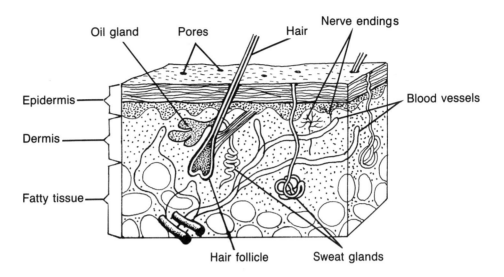

Hair is also part of the skin. It covers most of the body. Hair grows out of a small sac in the dermis called a **hair follicle**. Find the hair and the follicle in the picture.

Skin has two types of **glands**. Glands are parts of the body that **secrete**, or release, **fluids**. Find these two glands in the picture.

One type of gland is the **sebaceous gland**, or oil gland. It secretes oil into the hair follicle. This oil keeps the hair and skin healthy. When you brush your hair, you move the oil through the hair and make it shiny.

The other type of gland is the **sweat gland**. It secretes sweat made of water and salt. When the body becomes too hot from exercise or sunlight, the sweat glands go to work to cool you off. When you sweat, your body keeps an even temperature. Sweat leaves the body through small holes in the epidermis called **pores**. The sweat evaporates, or disappears, into the air and takes the heat with it. When this happens, you feel cooler. Find the pores in the picture. They are on the surface of the epidermis.

Skin protects the body. The skin protects the body by keeping its water in. When you peel a potato and leave it overnight, it dries out and will spoil quickly. If a potato is not peeled, it will not dry out or spoil for a long time. The skin of the potato protects the potato in the same way your skin protects your body.

Skin also protects the body by keeping **germs** and **bacteria** that are in the air out of the body. If you cut or scratch your skin, you must keep it very clean. Germs that cause infection or sickness can enter the body where the skin is broken or cut.

Skin also protects the body because it contains nerve endings. These nerve endings help you to feel things. They let you know what you are touching. They let you feel softness, heat, cold, and pain. If you accidently cut or scratch a part of your skin or touch something hot, the touch will bring pain. You will move your body away from whatever causes the pain. In this way your skin helps protect your body.

In this reading you learned about skin. You learned about the parts of the skin and how the skin protects the body. In the next reading you will learn about the muscles in the body.

SELF-EVALUATION 1

VOCABULARY TICKETS Read the vocabulary tickets with your teacher and the whole class. Are there still some words you do not understand? Write these words in a notebook. With a partner, write some example sentences using these new words. Talk about the meaning of these words with your classmates.

VOCABULARY CHECK Here are some important words from this reading. Do you understand all of these words? Circle the words you do not understand. Then find the words in the reading. Talk about the meaning of these words with your classmates. If you know all the words, continue to the Question Review.

bacteria	fluids	nerve endings
blood vessels	freckles	pigment
calluses	germs	pores
connective tissue	glands	sebaceous gland
dermis	hair follicle	secrete
elastic	moles	sweat gland
epidermis		

QUESTION REVIEW Go back to the questions on pages 43 and 44. Look at your answers. Work with a partner. Look at your partner's answers too. Are they the same as your answers? Help each other write the correct answers.

PRE-READING 2

FOCUS QUESTION Skim the reading on pages 47, 48, and 49 to find the answer to the question below. Underline the answer in your book. Write the answer below.

■ *What are the three kinds of muscles? Where in the body are they found?* _____

DETAIL QUESTIONS

LEARNING STRATEGY
☆ Reading selectively

Read "The Muscles" on pages 47, 48, and 49. Find the details. Underline the answers in your book. Write the answers below. As you read, write down on your vocabulary tickets any words you do not understand or cannot pronounce.

1. What is an organ? _____

2. Name three organs in the body. _____

3. Give an example of a voluntary muscle and an involuntary muscle.

4. Name two important muscles in the arm. _____

5. Explain what happens to the arm muscles when you bend your arm.

READING 2 ★

The Muscles

A **muscle** is a type of tissue needed in order to move. Muscles move bones. Muscles move blood and food through our bodies. There are three kinds of muscles in the human body. They are **smooth muscles**, **cardiac muscles**, and **skeletal muscles**. They are found in different parts of the body, and they each have special and different jobs to do.

Smooth muscles are only found in some **organs** of the body. An organ is a part inside of the body that has a specialized function. For example, the lungs are organs. The job of the lungs is to give oxygen to the blood and remove carbon dioxide from the body. The eyes are organs. Their job is to see. Smooth muscles are also found in the stomach, the bladder, and the small intestines. Look at the picture below. It shows some of the organs in the human body. These organs all have smooth muscles. They move automatically. You cannot control them. These muscles are also very elastic. They can **expand**, or stretch, to more than three times their relaxed size, like a rubber balloon.

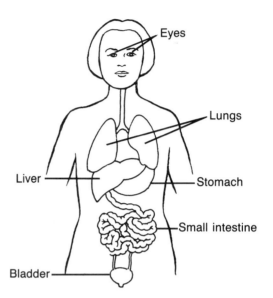

These are some of the organs with smooth muscles in the human body.

Cardiac muscles are another kind of muscle. They are found only in the heart. This kind of muscle makes the heart work. This muscle also works automatically. You cannot control this muscle. It moves in even, regular **contractions** all by itself. Muscles you cannot control are called **involuntary muscles**. Smooth muscles and cardiac muscles are involuntary muscles.

The third kind of muscle is called a skeletal muscle. Skeletal muscles move your body. They are attached to bones and let your body move where you want to move it. You have control over your skeletal muscles. They are not automatic. They are not involuntary. They move when and where you want them to move. They are called **voluntary muscles**.

To better understand the three kinds of muscles, study the picture below.

SMOULD MUSCLES
In organs
Involuntary

CARDIAC MUSCLES
In the heart only
Involuntary

SKELETAL MUSCLES
Attached to bones
Voluntary

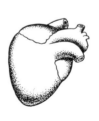 

Three kinds of muscles

Skeletal muscles relax and **contract**, or get tight, to move the body. For example, two skeletal muscles are at work when you bend your arm. These muscles in your arm are called the **biceps** and **triceps**. Both of them are connected to two bones—one at your shoulder and one below your elbow. When you want to bend your arm, your biceps contracts by getting tighter and shorter while the triceps relaxes. This pulls your arm up. When you want to straighten your arm again, your triceps gets tighter and shorter and your biceps relaxes. This pulls your arm back down. It is very easy to feel your biceps and triceps muscles at work.

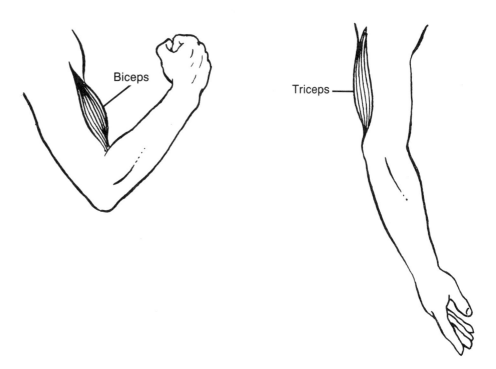

In this reading you learned about the muscles in the body. You learned that there are three kinds of muscles. You also learned where in the body these different muscles are found and how they work. Next, you will learn about another important part of the body—the skeletal system.

SELF-EVALUATION 2

VOCABULARY TICKETS Read the vocabulary tickets with your teacher and the whole class. Are there still some words you do not understand? Write these words in a notebook. With a partner, write some example sentences using these new words. Talk about the meaning of these words with your classmates.

VOCABULARY CHECK

Here are some important words from this reading. Do you understand all of these words? Circle the words you do not understand. Then find the words in the reading. Talk about the meaning of these words with your classmates. If you know all the words, continue to the Question Review.

biceps	muscle
cardiac muscles	organs
contract	skeletal muscles
contractions	smooth muscles
expand	triceps
involuntary muscles	voluntary muscles

QUESTION REVIEW

Go back to the questions on pages 46 and 47. Look at your answers. Work with a partner. Look at your partner's answers too. Are they the same as your answers? Help each other write the correct answers.

PRE-READING 3

FOCUS QUESTIONS

Skim the reading on pages 51, 52, 53, and 54 to find the answers to the questions below. Underline the answers in your book. Write the answers below.

■ *List five parts of the skeletal system.* _____

_____

■ *Why is the skeletal system important to the body?* _____

_____

DETAIL QUESTIONS

> **LEARNING STRATEGY**
> ☆ Reading selectively

Read "The Skeletal System" on pages 51, 52, 53, and 54. Find the details. Underline the answers in your book. Write the answers below. As you read, write down on your vocabulary tickets any words you do not understand or cannot pronounce.

1. Where can you find the smallest bones in the body? Where are the longest bones in the body? _____

_____

2. What is the hard, outer covering of the bone called? _____

_____

50 ★ CHAPTER 3

3. What is the job of the bone marrow? _____

_____

_____

4. Describe cartilage. Name two places it is found in the human body.

_____

_____

_____

5. Name two kinds of connective tissue. _____

_____

6. What are joints? _____

_____

READING 3 ★

The Skeletal System

The **skeletal system** is made up of bones, **bone marrow**, **cartilage**, **ligaments**, **tendons**, and **joints**. There are four main functions of the skeletal system. First, the skeletal system gives your muscles a place to hold on so that your body can move. Next, it supports your body. It holds it upright and gives it shape. The skeletal system also protects the body by covering the brain and all the important organs in a hard "shell." Finally, the skeletal system makes new blood cells.

Human adult bodies usually have 206 bones. Over half of your bones are in your feet and hands. You have 26 bones in each foot and 27 bones in each hand—106 in all. Where are the smallest bones in your body? The three smallest bones in the human body are inside your ears. These three bones together are about as big as your thumbnail.

The longest bone in the human body is called the femur. It is the long bone in the top part of the leg. A man who grows to be six feet tall may have a femur that is 20 inches long.

Look at the picture on the next page. The names of the biggest bones are given to you. It is not necessary to remember the names of all these bones. Instead, look at the picture and see if you can find two bones in the bottom of the leg, three bones in the arm and three bones in the head. Then, find three other bones and say where they are in the body.

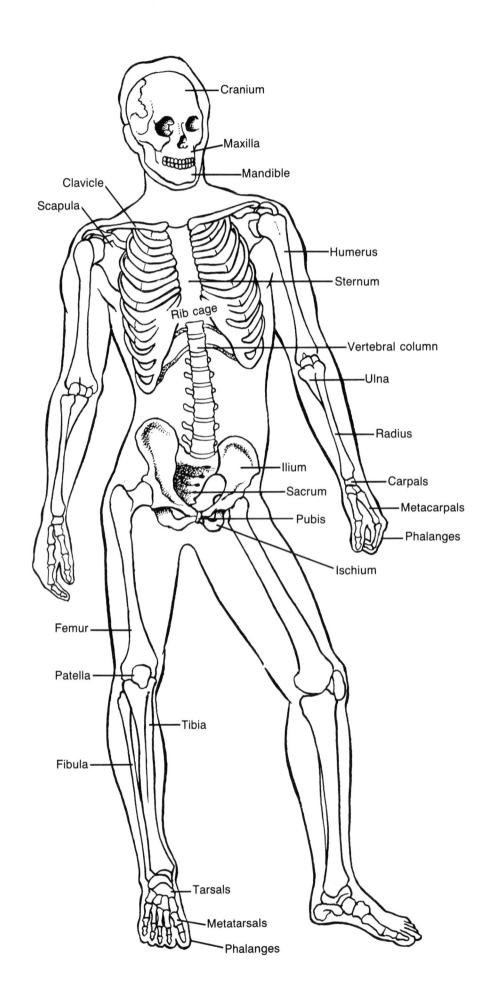

Bones have a hard, outer layer called the **periosteum** and a soft tissue on the inside called the **cortex**. The tissue inside the cortex is called bone marrow. The bone marrow forms new blood cells.

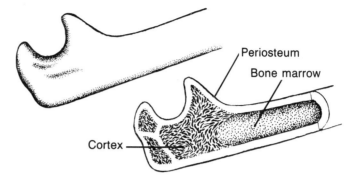

Our skeletal system is made up of more than bones. Some of the skeletal system is made of cartilage. This is especially true in young children. Cartilage is a **flexible** material that looks like bendable plastic. Cartilage is the material your outer ear and the end of your nose are made of. The skeletal system of a baby is mostly cartilage. This cartilage slowly changes to hard bone as a child grows. Because cartilage is bendable, it does not break easily. This is why children seldom get broken bones even when they play hard and fall a lot!

Connective tissue is also an important part of the skeletal system. Connective tissue looks like bands that hold two body parts together. Ligaments are one kind of connective tissue. Ligaments hold two bones together at a joint. Tendons are another kind of connective tissue. Tendons attach muscles to bones.

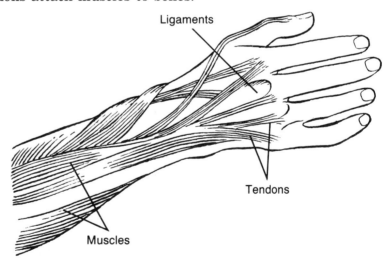

Bones are hard. They cannot bend. You bend your arms, legs, and other parts of your body at places where two bones come together. These places are called joints. Some joints do not move. For example, you have joints in your head that do not move. Some joints move just a little. The bones in the middle of your foot and in the middle of your hand move just a little. Some joints do not move, but they can stretch a little. Joints that stretch a little are in your **rib cage**. Find the rib cage in the picture

HUMAN ANATOMY ★ 53

on page 52. Some joints move a lot. The joints in your knees, hips, fingers, elbows, and wrists have a lot of movement.

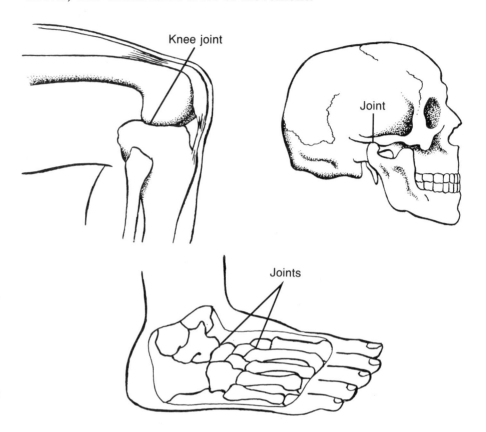

The joint in your knee moves a lot. The joints in your head do not move at all. The joints in your foot move a little.

In this reading you learned about the functions of the skeletal system. You also learned about the different parts of the skeletal system, how they work, and why they are important.

SELF-EVALUATION 3

VOCABULARY TICKETS Read the vocabulary tickets with your teacher and the whole class. Are there still some words you do not understand? Write these words in a notebook. With a partner, write some example sentences using these new words. Talk about the meaning of these words with your classmates.

VOCABULARY CHECK Here are some important words from this reading. Do you understand all of these words? Circle the words you do not understand. Then find the words in the reading. Talk about the meaning of these words with your classmates. If you know all the words, continue to the Question Review.

bone marrow	joints	rib cage
cartilage	ligaments	skeletal system
cortex	periosteum	tendons
flexible		

QUESTION REVIEW

Go back to the questions on pages 50 and 51. Look at your answers. Work with a partner. Look at your partner's answers too. Are they the same as your answers? Help each other write the correct answers.

CHAPTER REVIEW

Now that you have completed your reading about the skin, muscles, and bones of the human body, go back to pages 41, 42, and 43. Look at your first ideas about skin, muscles, and bones. Have your ideas changed? What have you learned? Talk about your ideas with the teacher and the whole class.

EXTENSION ACTIVITIES

A. FINGERPRINTS

Look at the pictures of fingerprints. These are the three main types of fingerprints. With a stamp pad and ink, take prints of your thumb and two fingers. Which type of print do you have? Are they all the same type? Compare your fingerprints with two or three other students. Are they the same? Do you have prints on your toes, too? Do dogs and cats have fingerprints? Find out and tell the class.

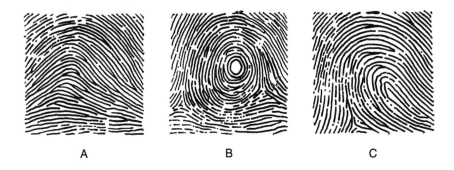

A   B   C

B. CHICKEN LEGS

LEARNING STRATEGIES
☆ Working cooperatively
☆ Taking notes

Materials

a chicken leg with the bottom and top of the leg joined together	a pair of scissors
a knife	tweezers
	two or three dishes

Sit down in small groups. Work together. Follow the instructions. Take turns. Talk about the parts of the chicken leg as you are making your observations. Choose someone to take notes for your group.

1. Look carefully at the skin. Can you see where the feathers were? Use your fingers and your scissors to take off the skin.

2. Look for the skeletal muscles. These muscles are attached to the bones. They are pink-gray in color. They make the leg move. Use your knife and carefully remove the skeletal muscles from the bone. Be careful not to break or cut the bones.

HUMAN ANATOMY ★ 55

3. Look for the connective tissue that holds the muscles to the underside of the skin. It looks like a thin, flat film that covers the muscles.
4. Look for tendons. Tendons attach muscles to bones. They look like shiny, white cords.

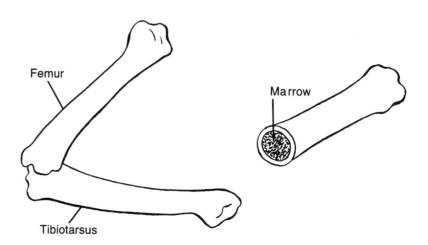

5. Remove all the muscles from the two bones and look for cartilage. Cartilage is white in color and is found at the end of long bones. It protects the ends of the two bones where they rub together. It is hard, but flexible. It is not as hard as bone. Look at the difference between the cartilage and the bone. Which is smooth and which is rough? Separate the two long bones. Scrape off the cartilage with your knife. Look at it carefully.
6. Compare the two bones. Are they the same or different? Measure how long they are. One bone is called the femur. The other bone is the tibiotarsus.

GLOSSARY

bacteria Microscopic living things that can cause sickness; germs.
biceps A muscle in the top front part of the arm.
blood vessels Small tubes that carry blood through the body.
bone marrow Tissue inside the bone where blood is made.
calluses Thick, hard places on the outside layer of skin formed from rubbing over a long period.
cardiac muscles Muscles found only in the heart.
cartilage Hard but flexible tissue.
connective tissue Thin, flat tissue that holds two body parts or tissues together.

contract To tighten up or get shorter.
contractions Regular movements.
cortex Soft tissue inside of the hard, outer layer of bone.
dermis The second layer of skin.
elastic Flexible or stretchable.
epidermis The outside layer of skin.
expand To get bigger.
flexible Bendable, moveable.
fluids Substances that flow like water, such as blood, tears, sweat.
freckles Small, brown spots on the skin.
germs Microscopic living things that can cause sickness; bacteria.

glands Parts of the body that make and secrete fluids such as sweat, oil, tears.

hair follicle A small sac in the skin from which hair grows.

human anatomy The study of the structure of the human body.

involuntary muscles Muscles that you cannot control; muscles that move automatically.

joints Places where two bones meet.

ligaments Connective tissue that holds two bones together at a joint.

moles Brown or black spots on the skin.

muscle A type of tissue needed in order to move.

nerve endings Tiny threads found near the surface of the skin that carry feelings such as hot and cold.

organs Inside body parts that have special functions, such as the heart or the lungs.

periosteum The hard, outer layer of bone.

pigment The material in skin that gives it color.

pores Small openings in the surface of the skin.

rib cage The bones that cover all of the organs of the chest.

sebaceous gland A skin gland that produces oil.

secrete To make and give off fluids.

skeletal muscles Muscles that attach to bones and move the body.

skeletal system The body system made up of bones, marrow, ligaments, tendons, cartilage, and joints.

smooth muscles Muscles of organs, for example, the stomach, the small intestine, and the bladder; involuntary muscles.

sweat gland A type of gland in the skin that makes and gives off sweat.

tendon Connective tissue that holds muscle to bone.

triceps A muscle in the top back part of the arm.

voluntary muscles Muscles that you can control and move.

CHAPTER 4

HUMAN PHYSIOLOGY
DIGESTION, RESPIRATION AND CIRCULATION

INTRODUCTION

Human physiology is the study of the different parts and systems of the human body and what they do. A human body has four basic needs—air, water, food, and **elimination**. A human body has three systems that take care of all these needs. They are the **digestive system**, the **respiratory system**, and the **circulatory system**. The digestive system takes in food. It changes food to a form the body can use. It takes away food that the body cannot use. The respiratory system brings **oxygen** into the body and takes **carbon dioxide** away from the body. The circulatory system carries food, water, and oxygen through the blood and carries away **waste** material. In this chapter you will learn about the most important parts of these three systems. You will learn how these systems function.

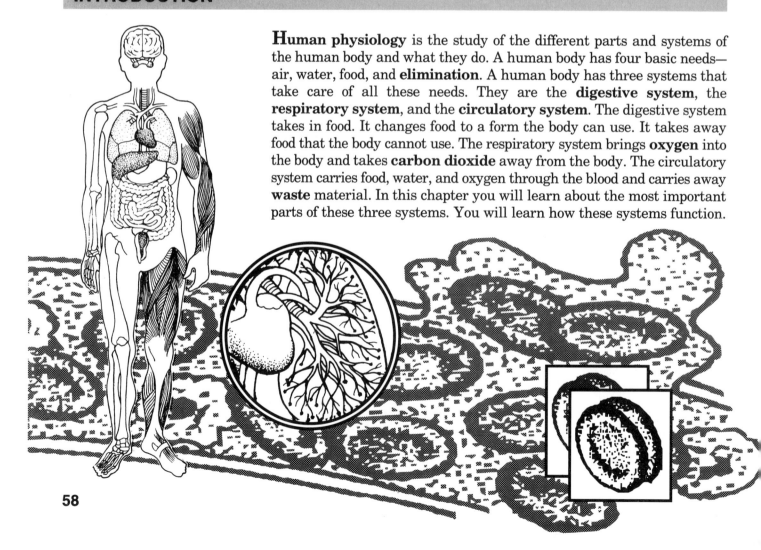

CRITICAL THINKING ACTIVITIES

WHAT DO YOU ALREADY KNOW ABOUT DIGESTION, RESPIRATION, AND CIRCULATION?

LEARNING STRATEGIES
☆ Using prior knowledge
☆ Working cooperatively

Read these sentences. Draw a circle around the words you do not understand. Underline the words you cannot pronounce.

You cannot live very long without food.
Food gives you energy to work and play.
It is necessary to chew your food before you swallow it.
Your heart is like a pump that moves blood around your body.
You can feel your heartbeat when you put your hand on your chest.
Blood moves through large and small tubes called blood vessels.
If you cut your skin, it will bleed.
Bleeding will stop automatically if the cut is small.

Sit down with a partner. Look at your book and your partner's book. Help each other understand the words that are circled. Help each other pronounce the words that are underlined.

THINK ABOUT THESE IDEAS

LEARNING STRATEGIES
☆ Inferencing
☆ Working cooperatively
☆ Self-evaluation

Work in groups of three or four. Work together to answer these questions. If you are not sure about your answers, guess!

1. Your teacher will pass around a jar of pickles. Smell them. When you smell the pickles (or when you think about food), do you feel water enter your mouth? Yes or no? What is the name of this "water"? Why does it enter your mouth?
2. Why does a small cut bleed and then just stop bleeding?
3. How long can you hold your breath? Find out who in your group can hold his or her breath the longest. Why can you not hold your breath as long as you want to?

When your group finishes talking about these ideas, share your ideas with the whole class. Are your ideas different? Are they similar? After you read this chapter, look at these ideas and your answers again. Do not worry if your answers are right or wrong.

GROUP OBSERVATIONS

LEARNING STRATEGIES
☆ Inferencing
☆ Working cooperatively
☆ Self-evaluation

Sit down in groups of three or four. Read the questions below and talk about what you see, feel, or hear. Remember, this is not a test. Write anything you can think of.

1. You can feel your heartbeat by putting your hand on your chest. You can also feel your heartbeat in large **blood vessels**, or tubes in the body called **arteries**. You have an artery in your wrist that is very near the surface of the skin. Put two or three fingers of

your right hand on the inside of your left wrist. (Do not use your thumb.) You can feel the blood being pumped through this artery. Count these heartbeats. Count your **pulse**.

2. Look at a watch and count how many times your heart beats in one minute. Write down the results for each group member. (It is usually between 60 and 150 beats per minute.) Now, all group members will stand up and run in place for two minutes. Count your pulse, or heartbeats, again. Write down the results for each student in your group.

Name	Pulse rate relaxed	Pulse rate exercised
_____	_____	_____
_____	_____	_____
_____	_____	_____
_____	_____	_____

Why does your heart beat faster during exercise? Do you have an idea?

3. Under your stomach is your small intestine. It is a long, coiled tube that the food you eat passes through.

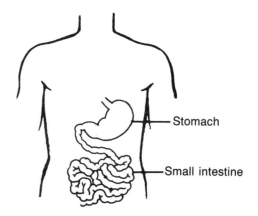

The small intestine is about 20 feet long. Take a ruler and measure in the room somewhere the distance of 20 feet. Now you can see exactly how long the small intestine is!

Share your observations with the whole class. After you finish reading this chapter, come back to these questions and observations and read them again. Are your answers the same?

PRE-READING 1

FOCUS QUESTION

Skim the reading on pages 62 and 63 to find the answer to the question below. Underline the answer in your book. Write the answer below.

■ *What is digestion?* _____

_____

DETAIL QUESTIONS

Read "Digestion" on pages 62 and 63. Find the details. Underline the answers in your book. Write the answers below. As you read, write down on your vocabulary tickets any words you do not understand or cannot pronounce.

LEARNING STRATEGY
☆ Reading selectively

1. What is mechanical digestion? _____

_____

2. What is chemical digestion? _____

_____

3. What is absorption? _____

_____

4. Why is saliva important? _____

_____

5. Name two digestive chemicals in the stomach. _____

_____

6. What happens to food in the small intestine? _____

_____

7. What happens to food in the large intestine? _____

_____

HUMAN PHYSIOLOGY

READING 1

Digestion

Food is necessary for survival. Everyone needs to eat. Many things happen to the food you eat once you put it into your mouth. Your body cannot use food when you first swallow it. The food must be changed. The process that changes food so that the body can use it is called **digestion**.

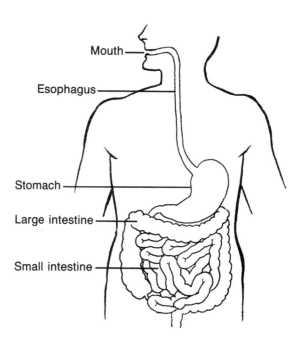

There are three steps to digestion—**mechanical digestion, chemical digestion**, and **absorption**. The food is first broken into small pieces by the teeth and then by the movement of the stomach muscles. This is called mechanical digestion. When you put food into your mouth, the teeth begin mechanical digestion by breaking up the food. The food is then mixed with **saliva**. Saliva is a kind of water that comes into your mouth. It has special chemicals to help digestion. Saliva will come into your mouth when you are ready to eat or when you just smell food or think about it! The breaking down of food by chemicals is called chemical digestion.

When you swallow food, it moves down a long tube called the **esophagus**. No digestion occurs in the esophagus. The food is there only for a short time.

The other end of the esophagus is connected to the stomach. Chemical digestion also occurs in the stomach. When food enters the stomach, the stomach muscles squeeze the food and move it around. The stomach also has special digestive chemicals called **acids** and enzymes that chemically digest the food. It takes about four hours for the food to be digested in the stomach. Chemical digestion breaks the food down into pieces so small that you would need a very powerful microscope to see them. Food is changed into a liquid form.

From the stomach, the digested food passes into a long, thin, coiled tube called the **small intestine**. This intestine is about 20 feet long. It is the most important digestive organ because food from the small intestine is absorbed directly into the blood and travels all around the body. This process is called absorption.

After the food is digested and absorbed into the blood, there is still some leftover food that cannot be absorbed. This **waste**, or unabsorbed food, along with some water pass into the **large intestine**. It stays in the large intestine for about one day. It is broken down further in the large intestine, and then it passes out of the body through the **anus**.

In this reading you learned about the digestive system, and that there are three steps in the digestive process. You also learned about what happens to the food in the human body from the time it enters your mouth until it passes out of the body through the anus.

SELF-EVALUATION 1

VOCABULARY TICKETS Read the vocabulary tickets with your teacher and the whole class. Are there still some words you do not understand? Write these words in a notebook. With a partner, write some example sentences using these new words. Talk about the meaning of these words with your classmates.

VOCABULARY CHECK Here are some important words from this reading. Do you understand all of these words? Circle the words you do not understand. Then find the words in the reading. Talk about the meaning of these words with your classmates. If you know all the words, continue to the Question Review.

absorption	digestion	mechanical digestion
acids	enzymes	saliva
anus	esophagus	small intestine
chemical digestion	large intestine	waste

QUESTION REVIEW Go back to the questions on page 61. Look at your answers. Work with a partner. Look at your partner's answers too. Are they the same as your answers? Help each other write the correct answers.

PRE-READING 2

FOCUS QUESTION

Skim the reading on pages 65 and 66 to find the answer to the question below. Underline the answer in your book. Write the answer below.

■ *What is respiration?* _____

_____

DETAIL QUESTIONS

Read "Respiration" on pages 65 and 66. Find the details. Underline the answers in your book. Write the answers below. As you read, write down on your vocabulary tickets any words you do not understand or cannot pronounce.

LEARNING STRATEGY
☆ Reading selectively

1. Finish these sentences: First, air enters through the _____ or _____. Then, it goes into the _____ and moves through the _____.

2. What is another name for the voice box? _____
_____

3. What is the job of the voice box? _____
_____

4. What is another name for the windpipe? _____
_____

5. What are the main organs of respiration? _____
_____

6. Finish this sentence: The two tubes that go into the lungs are called _____.

7. Which are larger, bronchioles or alveoli? _____
_____

8. What are the two gases that are exchanged in the respiration process? _____
_____

64 ★ CHAPTER 4

9. Where are the gases exchanged? _____

10. What is the diaphragm? _____

READING 2 ★

Respiration

Why do you breathe? You breathe to stay alive. When you breathe, you take air into your body. Air contains a gas called **oxygen**. You need oxygen for energy, and you need energy to keep warm, to build new tissue, and to move muscles.

When you breathe, you take in oxygen and let out **carbon dioxide**. The process of taking in oxygen and breathing out carbon dioxide is called **respiration**. Respiration, then, is an exchange of gases. Your body needs oxygen, but it does not need carbon dioxide.

How do you breathe? What happens when you take a breath of air? First, the air enters through the nose or mouth. Then it goes into the **pharynx** at the back of the throat and moves down through the **larynx**, or voice box. The larynx lets you speak. Next, the air continues through the **trachea**, or windpipe. The air then goes in two directions. It enters one of two tubes called **bronchi** that go into the **lungs**. The lungs are the two main organs of respiration. Both of the bronchi branch off into hundreds of small **bronchioles**. The bronchioles branch off into over 300 million tiny air sacs called **alveoli**.

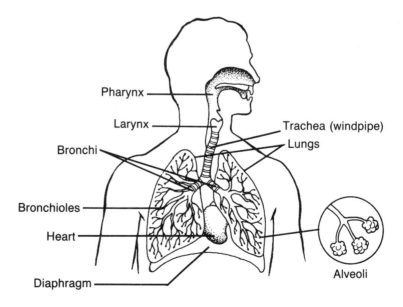

HUMAN PHYSIOLOGY ★ 65

Oxygen passes through the walls of the alveoli and into the smallest blood vessels, called **capillaries**. Blood vessels carry oxygen to all parts of the body. At the same time the blood is carrying oxygen, it is picking up carbon dioxide. It carries the carbon dioxide back to the capillaries. The carbon dioxide passes from the capillaries into the alveoli and into the lungs. You breathe out carbon dioxide.

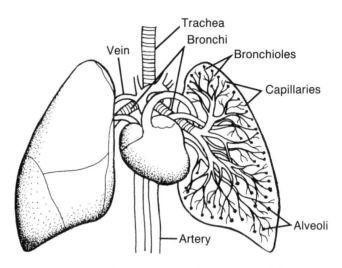

The blood picks up oxygen from the alveoli. The blood is pumped around the body by the heart. The blood returns to the lungs and leaves carbon dioxide waste in the alveoli. This carbon dioxide waste is breathed out of the body.

Your lungs have a big job to do. But they could not do this work without help from some muscles. One of these important muscles is the **diaphragm**. It is a large, elastic muscle under the lungs. The diaphragm moves up and down as you breathe. Other elastic parts of the lung help by stretching and squeezing to move the air in and out.

In this reading you learned some important things about the respiratory system. You learned why breathing is important for the human body, and what happens to the air that enters through your nose and mouth in the respiration process. In the next reading you will learn about another system in the human body, the circulatory system.

SELF-EVALUATION 2

VOCABULARY TICKETS   Read the vocabulary tickets with your teacher and the whole class. Are there still some words you do not understand? Write these words in a notebook. With a partner, write some example sentences using these new words. Talk about the meaning of these words with your classmates.

VOCABULARY CHECK

Here are some important words from this reading. Do you understand all of these words? Circle the words you do not understand. Then find the words in the reading. Talk about the meaning of these words with your classmates. If you know all the words, continue to the Question Review.

alveoli	carbon dioxide	oxygen
bronchi	diaphragm	pharynx
bronchioles	larynx	respiration
capillaries	lungs	trachea

QUESTION REVIEW

Go back to the questions on pages 64 and 65. Look at your answers. Work with a partner. Look at your partner's answers too. Are they the same as your answers? Help each other write the correct answers.

PRE-READING 3

FOCUS QUESTION

Skim the reading on pages 68, 69, and 70 to find the answer to the question below. Underline the answer in your book. Write the answer below.

■ *What is circulation?* _____

_____

DETAIL QUESTIONS

Read "Circulation" on pages 68, 69, and 70. Find the details. Underline the answers in your book. Write the answers below. As you read, write down on your vocabulary tickets any words you do not understand or cannot pronounce.

> **LEARNING STRATEGY**
> ☆ Reading selectively

1. Name three kinds of blood vessels. _____

_____

2. Finish these sentences: Blood leaves the heart through blood vessels called _____. Blood returns to the heart through blood vessels called _____.

3. Four things make up blood. One is liquid. Three are solid. Name the four things. _____

_____

HUMAN PHYSIOLOGY ★ 67

4. What is the function of white blood cells? _____

_____

5. What is the function of blood? Name five things the blood does.

_____

_____

READING 3 ★ ★ ★ ★ ★ ★ ★ ★ ★ ★ ★ ★ ★ ★ ★ ★ ★ ★ ★

Circulation

Circulation is the pumping of blood around the body. The **heart** pumps the blood through the body. The heart is the main organ in the circulatory system. The blood continually moves through the body and returns to the heart to be pumped out again. The heart pumps over 1,800 gallons of blood a day. When you listen to your heart, what do you hear? It sounds like it is saying, "lubb-dup, lubb-dup, lubb-dup." This sound is made by the closing of small "doors" or **valves** as the blood moves through the heart. The heart muscle never rests. It contracts and relaxes about 70 times a minute. It contracts and relaxes in a regular, even motion.

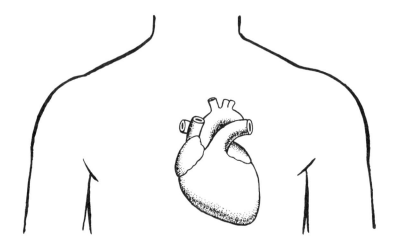

When blood travels through the body, it carries oxygen to all parts of the body. When you relax or when you sleep, your body does not need as much oxygen. Your heart can pump slowly when you are relaxed. When you work or play hard, your body needs a lot of oxygen. Your heart has to pump the blood very quickly to bring oxygen to the muscles and other parts of the body.

At rest your heart beats approximately 70 times per minute. Your heart beats approximately 120 times per minute when you exercise.

Tubes called blood vessels carry the blood to and away from the heart. There are three kinds of blood vessels. Blood moves out of the heart through blood vessels called **arteries**. It comes back to the heart through blood vessels called **veins**. The small blood vessels that connect the arteries and veins are called capillaries.

Your blood makes a trip around your body in about one minute. It makes over 1,000 trips through the body every day.

When blood is pumped out of the heart it goes into the largest arteries. The small arteries branch off into even smaller arteries. You can feel one of these arteries in your wrist when you take your **pulse**. A pulse is the regular movement of blood through the arteries caused by the contractions of the heart.

From the smaller arteries, blood enters the capillaries. Capillaries are the smallest blood vessels. You cannot see them without a microscope. Capillaries are like a bridge for the blood. Your blood leaves the arteries, crosses through the capillaries and goes into the veins. Then the blood travels back to the heart again. If you look at the surface of your skin on the inside of your arm, you can see veins.

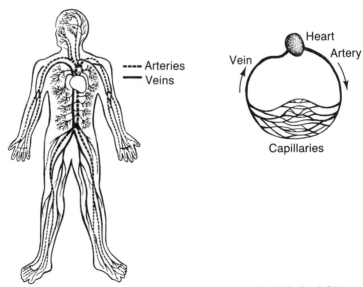

HUMAN PHYSIOLOGY ★ 69

What is blood made of? It has four main parts: **plasma**, **red blood cells**, **white blood cells**, and **platelets**. About half of your blood is made up of plasma, a pale yellow liquid. The other half is made up of red blood cells, white blood cells, and platelets. These are all solid materials.

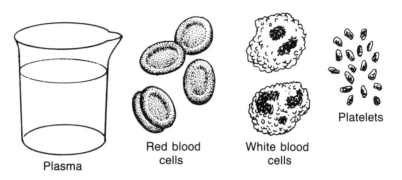

Each of these parts of the blood has a special job. Plasma is the carrier of waste products. It carries carbon dioxide waste to the lungs and other waste to other organs in the body. Red blood cells carry oxygen. White blood cells are the germ fighters. They help protect the body against disease. Platelets help you stop bleeding if you are cut or hurt. Platelets form **clots**. Platelets thicken the blood. They help to close the injury.

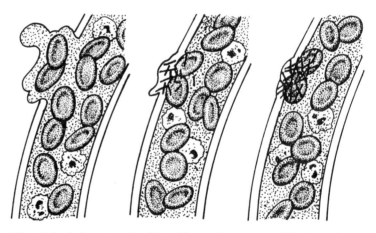

Platelets help stop the bleeding when your skin is cut. Platelets help the blood form clots.

In this reading you learned about the circulatory system and how it works. You read about the heart and how the blood moves around the body through the different blood vessels. You also learned a little about the blood and how it helps the body.

In this chapter, you learned about the three systems that help the body take care of its basic needs—digestion, respiration, and circulation.

SELF-EVALUATION 3

VOCABULARY TICKETS Read the vocabulary tickets with your teacher and the whole class. Are there still some words you do not understand? Write these words in a notebook. With a partner, write some example sentences using these new words. Talk about the meaning of these words with your classmates.

VOCABULARY CHECK Here are some important words from this reading. Do you understand all of these words? Circle the words you do not understand. Then find the words in the reading. Talk about the meaning of these words with your classmates. If you know all the words, continue to the Question Review.

arteries	plasma	valves
circulation	platelets	veins
clots	pulse	white blood cells
heart	red blood cells	

QUESTION REVIEW Go back to the questions on page 00. Look at your answers. Work with a partner. Look at your partner's answers too. Are they the same as your answers? Help each other write the correct answers.

CHAPTER REVIEW Now that you have completed your reading about human physiology, go back to pages 59, 60, and 61. Look at your first ideas about digestion, circulation, and respiration. Have your ideas changed? What have you learned? Talk about your ideas with the teacher and the whole class.

EXTENSION ACTIVITIES

A. EXPERIMENT WITH AIR IN THE LUNGS

Materials

| a large jar | a glass pan | a box of paper straws |

LEARNING STRATEGY
☆ Working cooperatively

How much air do you have in your lungs? How much air can you blow out?

1. First, fill the jar to the top with water.
2. Place the pan, upside down, on top of the glass.
3. Carefully flip the glass and pan over so that the glass is upside down in the pan.
4. Next, lift one side of the jar a little bit and put a straw under the edge.

5. Slowly blow out all the air you can blow. How much water leaves the jar? Mark the jar with a pen.
6. Take another straw. Have another student fill the jar and try the experiment.
7. Mark the jar again.

Have several of your classmates try. Who has the largest **lung capacity**? Who can blow out the most air? Some large adults can blow out a gallon of air. Of course in normal breathing, you breathe out much less than that.

B. CLASSIFYING HUMAN PHYSIOLOGY

LEARNING STRATEGY
☆ Grouping

Sit down with a partner. Look at the words in the chart. Decide which of the words are parts of the digestive system and put them in the correct column. Do the same for the circulation system and the respiratory system.

pulse	artery	large intestine
esophagus	enzymes	larynx
plasma	trachea	lungs
anus	platelets	capillaries
bronchi	saliva	clots
diaphragm	alveoli	acids

Digestion	Circulation	Respiration
_____	_____	_____
_____	_____	_____
_____	_____	_____
_____	_____	_____
_____	_____	_____
_____	_____	_____

C. VOCABULARY CLUES

Choose from the words below to complete the sentences that follow.

vessels   enzymes   waste   platelets   chemical

1. _____ are chemicals in the stomach that help in digestion.
2. Arteries and veins are two kinds of blood _____.
3. When the skin is cut, _____ help stop bleeding.
4. Carbon dioxide is a body _____. It must be removed.
5. There are two kinds of digestion—_____ and mechanical.

heart   trachea   alveoli   larynx   esophagus

6. The long tube from the mouth to the stomach is the _____.

7. The voice box is called the _____.

8. The tiny air sacs in the lungs where gases are exhanged are _____.

9. The main organ for the circulatory system is the _____.

10. The windpipe is sometimes called the _____.

veins   red   small intestine   digestive   diaphragm

11. The blood vessels that take blood back to the heart are called _____.

12. The system that processes food is called the _____ system.

13. The large muscle under the lungs is the _____.

14. The most important digestive organ is the _____.

15. Blood cells that carry oxygen are the _____ blood cells.

plasma   bronchi   arteries   capillaries   stomach

16. Your _____ stretches to hold about two to four quarts of food.

17. The smallest blood vessels in the body are called _____.

18. The two largest tubes in the lungs are _____.

19. The yellowish, liquid part of the blood is _____.

20. The _____ are the blood vessels that carry blood away from the heart.

HUMAN PHYSIOLOGY

GLOSSARY

absorption The taking of digested food into the blood stream from the small intestine.

acids Digestive chemicals in the stomach.

alveoli 300 million tiny air sacs connected to the bronchioles.

anus The opening in the body where digested food that will not be absorbed leaves the body.

arteries Blood vessels that carry blood away from the heart.

blood vessels Tubes that carry blood back and forth through the body.

bronchi Two large tubes that branch from the trachea into the lungs.

bronchioles Hundreds of small air tubes connected to the two large bronchi in the lungs.

capillaries The smallest blood vessels.

carbon dioxide A colorless gas breathed out as waste.

chemical digestion The breaking down of food by acids and enzymes in the digestive system.

circulation The movement of blood around the body.

circulatory system The parts of the body where blood flows to carry food, water, and oxygen.

clot A hardening or thickening of the blood where the skin is cut or broken.

diaphragm A large muscle under the lungs that moves up and down to help you breathe.

digestion The body process that changes food into a form that the body can absorb and use.

digestive system The parts of the body that change food into a form the body can use.

elimination The process of removing waste from the body.

enzymes Digestive chemicals in the mouth, stomach, and intestines.

esophagus A long tube for food that connects the pharynx to the stomach.

heart The organ that continually pumps blood around the body.

human physiology The study of the different parts and systems of the human body and what they do.

large intestine A part of the digestive system that holds and breaks down the waste food before it leaves the body.

larynx The voice box, or organ that makes sounds and speech possible.

lung capacity The amount of air the lungs can hold.

lungs The central respiratory or breathing organs.

mechanical digestion The breaking down of food by action of the teeth or the stomach muscles.

oxygen A colorless gas taken from the atmosphere by the lungs.

pharynx A tube at the back of the throat that both food and air pass through.

plasma A pale-yellow liquid part of blood that takes away waste.

platelets Small parts of the blood that help to heal a cut or injury.

pulse The regular movement of blood through the arteries caused by the contractions of the heart.

red blood cells Parts of blood that carry oxygen.

respiration Breathing in and out; the exchange of oxygen and carbon dioxide in the body.

respiratory system The parts of the body that allow people to breathe.

saliva The water or digestive fluid that is in the mouth.

small intestine A coiled tube connected to the stomach where food is digested and absorbed.

trachea The windpipe.

valves Small "doors" in the heart that open and close as blood is pumped through the heart.

veins Blood vessels that carry blood back to the heart.

waste Food that has been digested but not absorbed. It passes out of the body through the anus.

white blood cells Parts of the blood that fight germs and infection.

CHAPTER 5

HUMAN PHYSIOLOGY
THE SENSE ORGANS

INTRODUCTION

You learned in Chapter 4 that human physiology is the study of the different parts and systems of the human body and what they do. In this chapter, you will read more about human physiology. You will read about the **sense organs** in the human body. The sense organs are more commonly called the five senses. The sense organs are the ears, nose, eyes, tongue, and skin. The five senses are hearing, smelling, seeing, tasting, and touching. Your senses give information to your brain to help you move, control, and protect your body. In this chapter, you will learn more about the five senses and how they work in your body.

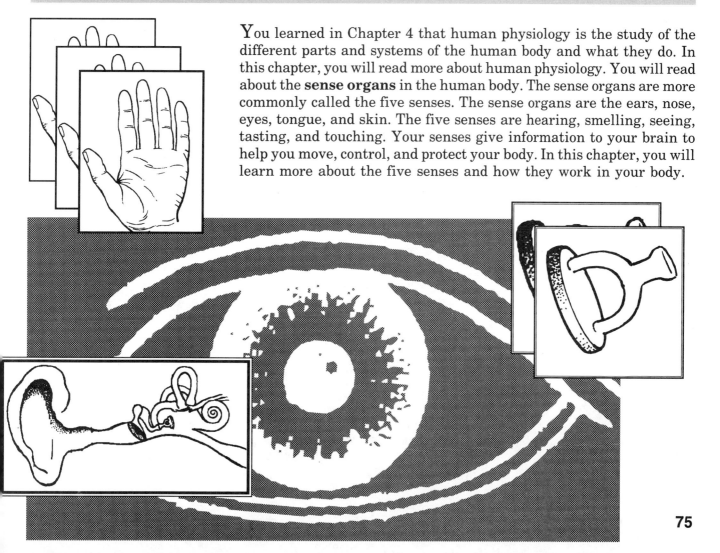

CRITICAL THINKING ACTIVITIES

WHAT DO YOU ALREADY KNOW ABOUT THE FIVE SENSES?

LEARNING STRATEGIES
☆ Using prior knowledge
☆ Working cooperatively

Read these sentences. Draw a circle around the words you do not understand. Underline the words you cannot pronounce.

We see with our eyes.
We hear with our ears.
We taste with our tongue.
We feel with our skin.
We smell with our nose.

Sit down with a partner. Look at your book and your partner's book. Help each other understand the words that are circled. Help each other pronounce the words that are underlined.

THINK ABOUT THESE IDEAS

LEARNING STRATEGIES
☆ Inferencing
☆ Taking notes
☆ Self-evaluation

Materials:

| a jar | liquid detergent with ammonia | paper cups | soda |

Work in groups of three or four. Work together to answer these questions. If you are not sure about your answers, guess!

1. Your teacher will stand in one corner of the classroom and open a jar. In the jar is something with a strong **odor**. Raise your hand as soon as you can smell the odor. Notice when your classmates raise their hands.

    a. Which classmates raise their hands first? Why?
    b. Can you think of one way the sense of smell protects you from danger?

2. Put a small amount of soda in a paper cup. Write down what you see, hear, smell, taste, and touch.

3. Have one person in the classroom whistle. You should be able to hear the whistle clearly. Now, send the person into another room and have him whistle again. You will no longer be able to hear the whistle clearly. Why?

4. Close your eyes. Be very quiet. Listen carefully to everything you hear for one minute. Try to remember what you hear. When your teacher calls time, write down five things you heard. Share your list with the other people in your group. Did you hear the same things? Did you hear something the other group members did not hear?

When your group finishes talking about these ideas, share your ideas with the whole class. Are your ideas different? Are they similar? After you read this chapter, look at these ideas and your answers again. Do not worry if your answers are right or wrong.

GROUP OBSERVATIONS

LEARNING STRATEGY
☆ Self-evaluation

Your teacher will give you three bags with one item in each bag. Put your hand in a bag and touch, smell, and taste the item. Do not look! Write the names of the things in the chart below. Give as much information about each one as you can. When you finish with one bag, pass the bag to another group member.

	Item	Smell	Taste	Feel
1.	_____	_____	_____	_____
2.	_____	_____	_____	_____
3.	_____	_____	_____	_____

Compare your answers with the other members in your group. Are they the same? Different? Share your answers with the whole class. After you finish reading this chapter, come back to these questions and observations and read them again. Are your answers the same?

PRE-READING 1

FOCUS QUESTION

Skim the reading on pages 78 and 79 to find the answer to the question below. Underline the answer in your book. Write the answer below.

■ *What causes sound?* _____

_____

DETAIL QUESTIONS

LEARNING STRATEGY
☆ Reading selectively

Read "The Sense of Hearing" on pages 78 and 79. Find the details. Underline the answers in your book. Write the answers below. As you read, write down on your vocabulary tickets any words you do not understand or cannot pronounce.

1. Name two parts of the ear. _____

_____

2. What happens to the eardrum when sound waves hit it?

_____

_____

HUMAN PHYSIOLOGY ★ 77

3. What are the three small bones in the ear called? _____

_____

4. How did they get their names? _____

_____

5. What presses on the hearing nerve cells? _____

_____

6. What do nerves inside the ear do? _____

_____

READING 1 ★ ★ ★ ★ ★ ★ ★ ★ ★ ★ ★ ★ ★ ★ ★ ★ ★ ★ ★

The Sense of Hearing

A bell rings, a baby cries, and a dog barks. Every day we hear thousands of sounds. Our world is full of sound. What causes sound? How do our **ears** let us hear sounds?

Sound is caused by **vibrations**, the quick back-and-forth movements of an object. The vibrations move through air, water, the ground, or some other substance. The vibrations move in waves. They are called **sound waves**. In order to understand how people hear sound waves, you must understand how the ear works.

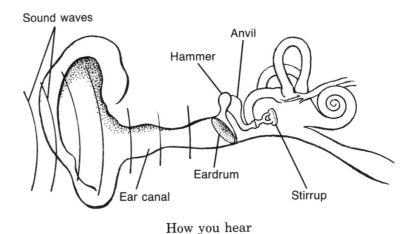

How you hear

There are three important parts to the ear: the **ear canal**, the **eardrum**, and the small bones. Each part is important for hearing. Sound waves enter the ear through the ear canal and hit the eardrum.

The eardrum is a thin skin that is stretched tightly across the inside of the ear. It is like the material that is stretched across the top of a drum. The eardrum begins to **vibrate**, or move back and forth quickly. This vibration causes three very small bones in the ear to vibrate. These little bones are called the **hammer**, **anvil**, and **stirrup**. They get their names because they look like these objects.

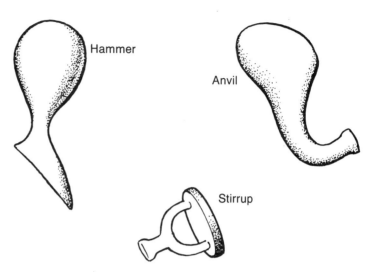

The small bones in the ear

These vibrations of the eardrum cause more vibrations in a liquid that fills the deepest part of the ear. The moving liquid presses on the hearing nerves. These nerves pass the sound message on to the **brain**. When the message reaches the brain, the person can hear the sounds.

It is important for humans to be able to hear sound. Sounds can warn of danger and emergencies. If you see a person cross the street into the path of an oncoming car, you would call to the person to watch out. The driver of the oncoming car would honk the horn to warn the person. Fire alarms warn people of fire. Sirens on ambulances and police cars tell you to move to the side. Some people cannot hear. They are **deaf** and cannot be warned of danger in the same way.

In this reading you learned about the sense of hearing. You learned about the ear and how humans hear. Next you will learn about two more senses. These are the senses of taste and smell.

SELF-EVALUATION 1

VOCABULARY TICKETS Read the vocabulary tickets with your teacher and the whole class. Are there still some words you do not understand? Write these words in a notebook. With a partner, write some example sentences using these new words. Talk about the meaning of these words with your classmates.

VOCABULARY CHECK

Here are some important words from this reading. Do you understand all of these words? Circle the words you do not understand. Then find the words in the reading. Talk about the meaning of these words with your classmates. If you know all the words, continue to the Question Review.

anvil	eardrum	stirrup
brain	ears	vibrate
deaf	hammer	vibrations
ear canal	sound waves	

QUESTION REVIEW

Go back to the questions on pages 77 and 78. Look at your answers. Work with a partner. Look at your partner's answers too. Are they the same as your answers? Help each other write the correct answers.

PRE-READING 2

FOCUS QUESTION

Skim the reading on pages 81 and 82 to find the answer to the question below. Underline the answer in your book. Write the answer below.

■ *What sense organs do people use for taste and smell?*

_____

_____

DETAIL QUESTIONS

LEARNING STRATEGY
☆ **Reading selectively**

Read "The Senses of Taste and Smell" on pages 81 and 82. Find the details. Underline the answers in your book. Write the answers below. As you read, write down on your vocabulary tickets any words you do not understand or cannot pronounce.

1. What are the bumps on the tongue called? _____

   _____

2. What is located inside the bumps on your tongue? _____

   _____

3. Where are sweet things tasted? _____

   _____

80 ★ CHAPTER 5

4. What is an odor? _____

_____

5. Why do people sniff? _____

_____

READING 2 ★

The Senses of Taste and Smell

Why does a potato chip taste salty? Why does sugar taste sweet? There are two **sense organs** you use to taste. One of these sense organs is the **tongue**. If you look in the mirror and stick out your tongue, you will see little bumps on it. These bumps are called **papillae**. Inside each of these bumps are tiny **taste buds**. Taste buds are cells that are connected to nerves. The nerves carry messages about the food you eat to the brain. The nerves tell your brain how something tastes. You can taste if something is **bitter, sour, sweet,** or **salty**. Look at the picture below to see where the taste buds are located and the different tastes you experience.

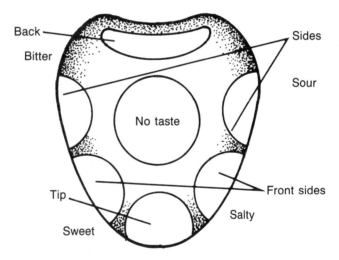

The taste areas of the tongue

You taste bitter things at the back of your tongue, sour and salty things on the sides, and sweet things on the tip. The tongue is only one part of the sense of tasting.

The other sense organ you use to taste is your **nose**. The nose is also the sense organ you use to smell. The smell of food plays a big part in how food tastes. If food smells good, it usually tastes good! Sometimes when you have a cold and your nose is stopped up, you cannot smell anything. When this happens, nothing you eat will taste very good either.

Everything that has a smell gives off a small amount of gas. This gas is called an **odor**. When you breathe in, the odor enters your nose. Some things have a weak odor. When things have a weak odor, you have to **sniff** to bring the odor into your nose. There are special nerves in the nose that send the "smell message" to the brain. The picture below shows how the sense of smell works.

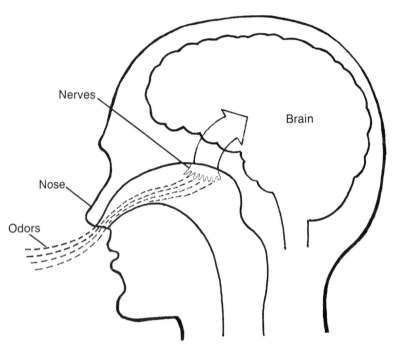

Odor enters through the nose and passes to the nerves. The nerves send a "smell message" to the brain.

It is important to be able to smell things. Your sense of smell protects you from danger. You smell smoke when there is a fire. Food begins to smell bad when it is no longer good to eat. Animals such as skunks spray a liquid that has a bad odor to protect them from danger.

In this reading you learned about the senses of taste and smell and their two sense organs, the tongue and the nose. You also learned why these two senses are important. Next you will learn about another sense organ and the sense of sight.

SELF-EVALUATION 2

VOCABULARY TICKETS — Read the vocabulary tickets with your teacher and the whole class. Are there still some words you do not understand? Write these words in a notebook. With a partner, write some example sentences using these new words. Talk about the meaning of these words with your classmates.

VOCABULARY CHECK Here are some important words from this reading. Do you understand all of these words? Circle the words you do not understand. Then find the words in the reading. Talk about the meaning of these words with your classmates. If you know all the words, continue to the Question Review.

bitter sniff
nose sour
odor sweet
papillae taste buds
salty tongue
sense organs

QUESTION REVIEW Go back to the questions on pages 80 and 81. Look at your answers. Work with a partner. Look at your partner's answers too. Are they the same as your answers? Help each other write the correct answers.

PRE-READING 3

FOCUS QUESTION Skim the reading on pages 84 and 85 to find the answer to the question below. Underline the answer in your book. Write the answer below.

■ *Name the main parts of the eye.* _____

_____

DETAIL QUESTIONS Read "The Sense of Sight" on pages 84 and 85. Find the details. Underline the answers in your book. Write the answers below. As you read, write down on your vocabulary tickets any words you do not understand or cannot pronounce.

LEARNING STRATEGY
☆ **Reading selectively**

1. What is the function of the iris? _____

_____

2. When does the pupil change size? _____

_____

3. Name two important functions of the eyelid. _____

_____

HUMAN PHYSIOLOGY

4. On which part of the eye does a picture form? _____

5. How does the brain help you to see? _____

6. Name a common eye problem. _____

7. How can eye problems be corrected? _____

READING 3 ★ ★ ★ ★ ★ ★ ★ ★ ★ ★ ★ ★ ★ ★ ★ ★ ★ ★ ★

The Sense of Sight

The **eye** is the sense organ of sight. You see with your eyes. Your eyes work like a very good camera. They can take pictures that are still or moving, in color or in black and white, and from a distance or close up. Of course, your eyes are better than a camera! In this reading you will learn how your eyes work and how you see.

The eye is made up of different parts: the **iris**, **pupil**, **eyelid**, and **retina**. The picture below shows the different parts of an eye.

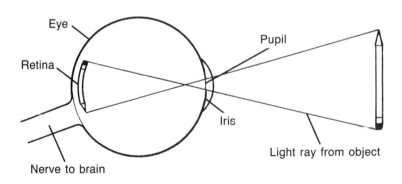

The iris is a muscle. It is the part of the eye that lets in the right amount of light. The big, colored circle in the center of the eye is the iris. **Pigment** gives the iris its color. The color of the iris is different in different people. Look at your classmates' eyes. What color irises do you see?

In the center of the iris there is a hole that lets in the light. This hole is the pupil. The iris muscle can change or adjust the size of the pupil. The pupil will **enlarge** if the light is dim and get smaller when the light is bright.

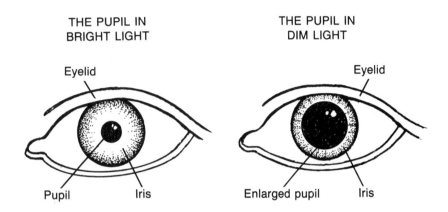

The eyelid is another important part of the eye. It has two important functions. The eyelid controls the amount of light that enters the eye. When you want to keep out light, you can lower your eyelid. Also, raising and lowering the eyelids helps keep the eyes moist.

Another important part of the eye is the retina. The retina is the part of the eye that receives the image and focuses the light. A picture forms on the retina in the back of the eye. The image on the retina is upside down. How does the image get right side up, so you see normally?

Light enters through the pupil in the eye and is received by the nerves in the retina. When the nerves in the retina receive the light, they send a "picture message" to the brain. This picture message is upside down. The brain changes the message into a right-side-up picture. The brain performs a very important function in the sense of sight.

Sometimes people need glasses because they cannot focus the light properly. The picture is not clear. Three of the most common eye problems are being **nearsighted**, **farsighted**, or having an **astigmatism**. If you are nearsighted, you can see things clearly only if they are very near. If you are farsighted, you can see things clearly only if they are far away. If you have an astigmatism, things look blurry whether they are near or far. All three problems can be corrected with eyeglasses or contact lenses. Eyeglasses or contact lenses help focus the light properly so that you can see clearly all the time.

In this reading you learned about the sense of sight. Next you will read about the sense of touch.

SELF-EVALUATION 3

VOCABULARY TICKETS   Read the vocabulary tickets with your teacher and the whole class. Are there still some words you do not understand? Write these words in a notebook. With a partner, write some example sentences using these new words. Talk about the meaning of these words with your classmates.

VOCABULARY CHECK   Here are some important words from this reading. Do you understand all of these words? Circle the words you do not understand. Then find the words in the reading. Talk about the meaning of these words with your classmates. If you know all the words, continue to the Question Review.

astigmatism	farsighted	pigment
enlarge	iris	pupil
eye	nearsighted	retina
eyelid		

QUESTION REVIEW   Go back to the questions on page 83 and 84. Look at your answers. Work with a partner. Look at your partner's answers too. Are they the same as your answers? Help each other write the correct answers.

PRE-READING 4

FOCUS QUESTION   Skim the reading on pages 87 and 88 to find the answer to the question below. Underline the answer in your book. Write the answer below.

■ *What is the job of the sensory nerves in the skin?* _____

_____

DETAIL QUESTIONS   Read "The Sense of Touch" on pages 87 and 88. Find the details. Underline the answers in your book. Write the answers below. As you read, write down on your vocabulary tickets any words you do not understand or cannot pronounce.

> **LEARNING STRATEGY**
> ☆ Reading selectively

1. Name five different messages the brain can receive from the nerves in the skin. _____

_____

2. Name one place that sensory nerves can be found. _____

_____

3. What makes some parts of your body more sensitive to touch than other parts? _____

4. Why is the sense of touch important? _____

READING 4 ★

The Sense of Touch

The **skin** is the sense organ of touch. It has millions of nerves that make it sensitive to touch. When something touches the skin, the **sensory nerves** in the skin send messages to the brain. The brain receives the messages as unpleasant or pleasant feelings. There are five kinds of messages the brain can receive from the sensory nerves in the skin. These messages are **pain**, heat, cold, **pressure**, and light touch.

Different kinds of sensory nerves are found all over your body. Some sensory nerves are found near the hairs on your body, others in hairless areas, while still others are found deeper inside your body. Some parts of your body are more sensitive to touch, or can feel things more strongly, than other parts. This is because these parts, like the tip of your tongue and tips of your fingers, have more sensory nerves. What other places on your body are more sensitive to touch? What places are the least sensitive to touch?

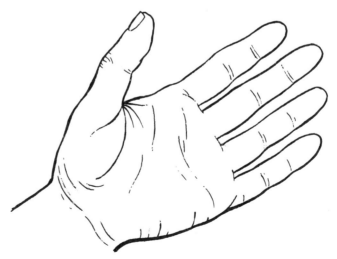

There are many nerves in the hand and fingers that are sensitive to different kinds of touch. Where is your hand the most sensitive?

HUMAN PHYSIOLOGY ★ 87

The sense of touch is very important. You can learn about your body through the sense of touch. Babies learn much of what they know about the world through their sense of touch. If a baby touches a hot stove, the nerves in the skin send a message of pain and heat to the brain. Almost immediately, the baby will remove his or her hand. If babies did not have this warning signal, they could seriously injure themselves.

When you are injured, you often wish you could not feel pain; yet, it is important that you feel pain. Pain protects you and lets you know if there is something wrong in your body.

In this chapter you learned about the five senses and the sense organs.

SELF-EVALUATION 4

VOCABULARY TICKETS Read the vocabulary tickets with your teacher and the whole class. Are there still some words you do not understand? Write these words in a notebook. With a partner, write some example sentences using these new words. Talk about the meaning of these words with your classmates.

VOCABULARY CHECK Here are some important words from this reading. Do you understand all of these words? Circle the words you do not understand. Then find the words in the reading. Talk about the meaning of these words with your classmates. If you know all the words, continue to the Question Review.

pain   sensory nerves
pressure   skin

QUESTION REVIEW Go back to the questions on pages 86 and 87. Look at your answers. Work with a partner. Look at your partner's answers too. Are they the same as your answers? Help each other write the correct answers.

CHAPTER REVIEW Now that you have completed your reading about the five senses, go back to pages 76 and 77. Look at your first ideas about the five senses. Have your ideas changed? What have you learned? Talk about your ideas with the teacher and the whole class.

EXTENSION ACTIVITIES

A. OPTICAL ILLUSIONS

LEARNING STRATEGY
☆ Imagery

Everyday experience teaches us that light travels in a straight line. Our eyes adjust so we can touch the things we see. For example, if you close one eye and try to pick up a pin, it will be more difficult than if you use both your eyes. The balance is easily upset.

The picture below shows examples of optical illusions. Your eyes see things that are not true. If you measure the figures, you will see that they will not support what you first saw.

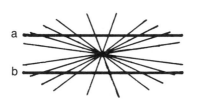

A. Are the horizontal lines parallel?

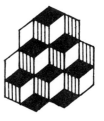

B. Count the cubes and count them again.

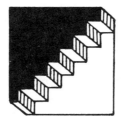

C. Look steadily at the staircase. Then slowly turn the book so that the staircase is upside down.

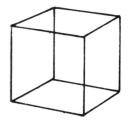

D. Are you looking at the top or bottom of the cube?

B. REVERSED HANDWRITING

Produce reversed writing by placing a piece of carbon paper, carbon side up, under a sheet of plain paper. Write something on the paper and you will have reversed writing on the other side. Read the reversed writing by holding it in front of a mirror. Look in the mirror while you write something. Watch the pencil.

C. VIBRATING RULER

Place a ruler on a table so that about two-thirds to three-quarters of it sticks out from the table edge. Hold down one end on the table. Bend the other end and let go quickly. The ruler should vibrate up and down. Listen to the sound you hear. Repeat the experiment several times, each time with less of the ruler sticking out. What differences do you hear in the sound the ruler makes?

D. INSTRUMENT OF TOUCH

Materials

| two straight pins   a pencil   masking tape |

Make a simple instrument of touch. Place two straight pins about one inch apart on a short piece of masking tape as in the picture below.

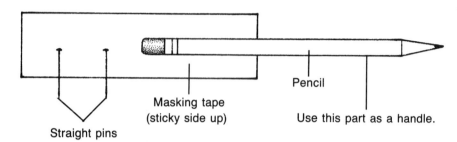

Place a pencil along the tape as shown. Cover with a second piece of masking tape. When you have the finished instrument, get a partner to help you test it. Your partner's eyes should be closed and both hands should be behind his or her back. Touch your partner's hand and ask if he or she feels one pin or two. Continue testing by gradually moving up the arm. Where is the skin most sensitive?

E. HOLD YOUR NOSE!

LEARNING STRATEGIES
- ☆ Inferencing
- ☆ Working cooperatively

In this chapter, you learned that the senses of taste and smell are closely related. Try an experiment to see if this is true. You and your teacher will bring different foods to class such as carrots, popcorn, celery, cookies, etc. You will have a chance to eat some of these foods. Close your eyes so you cannot see what you are eating and hold your nose. Your partner will give you a small taste of one of the sample foods. Try to guess what the food is. Your partner will write down your guesses.

GLOSSARY

anvil A small bone in the ear that looks like an anvil.
astigmatism A common problem with eyesight that causes objects to appear blurred or out of focus.
bitter The taste area in the back of the tongue.
brain The organ that controls mental and physical actions; located in the head.
deaf Not able to hear.
ear canal The narrow passage to the eardrum.

eardrum A thin skin that is stretched tightly across the inside of the ear.
ears Sense organs for hearing.
enlarge To get bigger.
eye The sense organ for sight.
eyelid A covering over the eye that blinks and closes.
farsighted A common problem with eyesight; things are seen best at a long distance.

hammer A small bone in the ear shaped like a hammer.

iris A muscle in the eye that controls the size of the pupil to let the correct amount of light into the eye.

nearsighted A common problem with eyesight; things are seen best close up.

nose The sense organ for smell.

odor A small amount of gas in the air that is sensed with the nose.

pain A feeling of hurt, suffering, or distress.

papillae Small bumps on the tongue for tasting things.

pigment Color.

pressure A firm touch.

pupil A small hole in the middle of the iris; it regulates the light that comes into the eye.

retina The back part of the eye that receives the "picture" or image.

salty A taste area on the sides of the front of the tongue.

sense organs The nose, eyes, ears, skin, tongue.

sensory nerves Special tissue that sends messages from the sense organ to the brain.

skin The sense organ for touch.

sniff To quickly bring air into the nose.

sound waves Movement of air, water, the ground, or some other substance.

sour The taste area on the sides of the back of the tongue.

stirrup A small bone in the ear that is shaped like a stirrup.

sweet The taste area on the front of the tongue.

taste buds Cells in the papillae that detect flavor or taste.

tongue The sense organ for taste.

vibrate To move quickly back and forth.

vibrations Quick back-and-forth movements.

CHAPTER 6

HUMAN ECOLOGY
HEALTHFUL LIVING: NUTRITION AND EXERCISE

INTRODUCTION

Human ecology is the study of human beings in the world they inhabit. In this chapter, you will read about the things people need to do to be healthy. A healthy person eats the right kinds of foods and gets enough exercise. You will learn about the things you need to eat and the kinds of exercise you need to do in order to be healthy.

CRITICAL THINKING ACTIVITIES

WHAT DO YOU ALREADY KNOW ABOUT HEALTHFUL LIVING?

LEARNING STRATEGIES
☆ Using prior knowledge
☆ Working cooperatively

Read these sentences. Draw a circle around the words you do not understand. Underline the words you cannot pronounce.

> Exercise will burn up calories.
> Exercise improves muscle tone.
> Vitamins and minerals keep the body healthy.
> A well-balanced diet is important for good health.
> It is important to get the right amount of exercise.

Sit down with a partner. Look at your book and your partner's book. Help each other understand the words that are circled. Help each other pronounce the words that are underlined.

THINK ABOUT THESE IDEAS

LEARNING STRATEGIES
☆ Inferencing
☆ Self-evaluation

Work in groups of three or four. Work together to answer these questions. If you are not sure about your answers, guess!

1. Who should exercise?
2. What kind of exercise do you think is best for you?
3. How often should you exercise?
4. What kinds of foods do you need to eat in order to be healthy?

When your group finishes talking about these ideas, share your ideas with the whole class. Are your ideas different? Are they similar? After you read this chapter, look at these ideas and your answers again. Do not worry if your answers are right or wrong.

GROUP OBSERVATIONS

LEARNING STRATEGIES
☆ Inferencing
☆ Taking notes
☆ Working cooperatively

1. Record the following observations in a notebook.

    a. Make a list of all the food you have eaten since yesterday at this time. Try to remember everything! Then add a list of the food you plan to eat before you go to bed tonight.
    b. Which foods on your list are good for you? Explain why.
    c. Which foods on your list are not good for you? Explain why.
    d. Is your diet balanced? Explain why or why not.

2. Share the list of foods you have eaten with two other people in your class. Get their answers to the following questions. Write their responses in your notebook.

    a. Do you think I have eaten food that is good for me? Why or why not?
    b. Do you think my diet was balanced? Why or why not?

3. Now, write down two different things you can do to improve your diet.
4. Work together in groups of three or four. Write your responses in a notebook.

   a. Make a list of eight different things you can do to get exercise.
   b. Decide on the three forms of exercise that your group thinks are the most helpful and explain why.
   c. Name three forms of exercise you prefer to do with a group.
   d. Name the forms of exercise that your group likes the most. Be prepared to share your answers with the whole class.

After you finish reading this chapter, come back to these questions and observations and read them again. Are your answers the same?

PRE-READING 1

FOCUS QUESTION

Skim the reading on pages 95 and 96 to find the answer to the question below. Underline the answer in your book. Write the answer below.

■ *What kind of food does your body need to be healthy?*

_____

_____

DETAIL QUESTIONS

LEARNING STRATEGY
☆ Reading selectively

Read "Nutrients and Healthful Eating" on pages 95 and 96. Find the details. Underline the answers in your book. Write the answers below. As you read, write down on your vocabulary tickets any words you do not understand or cannot pronounce.

1. What are three ways that nutrients help the body? _____

_____

2. Name three classes of nutrients. _____

_____

3. What nutrient is an important source of energy? _____

_____

4. Which nutrients help prevent disease? _____

_____

5. What nutrient is essential? _____

_____

6. Is there any food that has all the nutrients you need every day?

_____

READING 1 ★ ★ ★ ★ ★ ★ ★ ★ ★ ★ ★ ★ ★ ★ ★ ★ ★ ★ ★

Nutrients and Healthful Eating

What have you eaten today? A candy bar? Cold cereal? A salad? An apple? A hamburger? It is important to eat food that is good for you. Eating the right food is important for good health.

The body needs food that has **nutrients**. Nutrients make food healthful. Nutrients give your body strength and promote good health. Nutrients help the body in three important ways. They promote growth, replace old cells, and provide energy. The body needs energy. The energy in food is measured in units called **calories**. All food has calories. You must eat food with enough calories so that you will have enough energy. If you eat food with too many calories, you will not use the calories. Your body will store the energy and you will get fatter.

There are six classes of nutrients: **carbohydrates**, **fats**, **proteins**, **vitamins**, **minerals**, and water. Carbohydrates are sugars and starches. They are found in potatoes, bread, cereal, and spaghetti. They are the most important source of energy for the body. Runners, swimmers, and other athletes eat foods high in carbohydrates before a race or competition in order to have enough energy. Fats are another source of energy. They are stored under the surface of the skin. When the body needs extra energy, it uses the stored fat. Butter, red meat, and cream are high in fats.

Proteins are important for growth. The most common source of protein is meat. Protein can also be found in other foods such as nuts and beans. Vitamins are important for growth and the prevention of disease. You can get vitamins from food, or you can take vitamins in the form of pills. Minerals help keep the many different parts of the body functioning. Calcium, magnesium, and iron are examples of minerals that are important for your body to function properly. Water is an **essential** nutrient for your body. You must have water every day in order to function properly.

There are many different foods to choose from in the supermarket. How do you know if you are choosing food that will give you the right amount of nutrients? Will a chocolate cake, a candy bar, a hamburger, or french fries give you the nutrients you need to be healthy? Do you have to eat food you do not like in order to be healthy?

No single food has all the nutrients you need every day. It is important to eat many different kinds of food. You can find food you like and that is good for you, too!

In this reading you learned about nutrients in the food you eat, and why they are important to your body. Next you will learn about how you can get the nutrients you need by choosing the right kind of food.

SELF-EVALUATION 1

VOCABULARY TICKETS Read the vocabulary tickets with your teacher and the whole class. Are there still some words you do not understand? Write these words in a notebook. With a partner, write some example sentences using these new words. Talk about the meaning of these words with your classmates.

VOCABULARY CHECK Here are some important words from this reading. Do you understand all of these words? Circle the words you do not understand. Then find the words in the reading. Talk about the meaning of these words with your classmates. If you know all the words, continue to the Question Review.

    calories   minerals
    carbohydrates nutrients
    essential   proteins
    fats   vitamins

QUESTION REVIEW Go back to the questions on pages 94 and 95. Look at your answers. Work with a partner. Look at your partner's answers too. Are they the same as your answers? Help each other write the correct answers.

PRE-READING 2

FOCUS QUESTION Skim the reading on pages 97, 98, and 99 to find the answer to the question below. Underline the answer in the reading. Write the answer below.

■ *Why is it important to choose food from the basic food groups?*

_____

_____

DETAIL QUESTIONS

LEARNING STRATEGY
☆ Reading selectively

Read "The Basic Food Groups" on pages 97, 98, and 99. Find the details. Underline the answers in your book. Write the answers below. As you read, write down on your vocabulary tickets any words you do not understand or cannot pronounce.

1. What are the four basic food groups? _____
   _____

2. What are dairy foods? _____
   _____

3. Name one dairy food. _____

4. Why are dairy foods important to the body? _____
   _____

5. What are legumes? _____
   _____

6. To which food group does tofu belong? _____
   _____

7. Name one food high in fiber. _____

8. Why is fiber important to the body? _____
   _____

9. What are examples of cereal grains? _____
   _____

10. Finish this sentence: Eating the right food in the right amounts

    is called a _____.

READING 2 ★

The Basic Food Groups

In the last reading, you learned that you can find food you like and that is good for you, too. There are four important groups of food. By choosing foods from these four groups, you can get the nutrients your body needs to be healthy. The four basic food groups are the **dairy** group, the protein group, the fruit and vegetable group, and the bread and cereal group. You need food from each of these food groups every day. A certain number of average servings are recommended from each group each day. This amount will vary depending on the age of the person. Children sometimes need more servings than adults in some food groups.

A daily guide to good eating

DAIRY GROUP The food in this group includes milk and all foods that are made from milk such as cheese, yogurt, and ice cream. Foods in this group are important sources of minerals like calcium and phosphorus, of protein, and of vitamins A and D. You need at least two average servings from this group every day.

PROTEIN GROUP Beef, pork, fish, chicken, and eggs belong to this group. There are other important protein sources such as **legumes** (dried beans and peas), nuts, peanuts, peanut butter, and all products made from soybeans, such as **tofu**. Tofu is thickened soymilk similar to cheese. You should have at least two average servings from this group each day.

FRUIT AND VEGETABLE GROUP Vegetables are important sources for minerals, vitamins, and **fiber**. Fiber is rough, raw, or coarse food such as raw vegetables. It is important to have fiber in your **diet**. Fiber helps your digestive system run smoothly. Fruits and vegetables are important sources of fiber for your body. You should have four average servings from the fruit and vegetable group each day.

BREAD AND CEREAL GROUP Food in this group comes from plants called **cereal grains**, such as oats, rye, corn, and wheat. Breads, oatmeal, and bran are all foods from this group. They also supply important vitamins and minerals as well as roughage for your body. You should have three average servings from this group every day.

If you want to keep your body healthy, choose food from the basic food groups. Healthful eating means eating the right food in the right amounts. This is called a **balanced diet**.

In this reading you learned about the four basic food groups. You also learned what foods you need for a balanced diet. Next you will read about exercise.

SELF-EVALUATION 2

VOCABULARY TICKETS Read the vocabulary tickets with your teacher and the whole class. Are there still some words you do not understand? Write these words in a notebook. With a partner, write some example sentences using these new words. Talk about the meaning of these words with your classmates.

VOCABULARY CHECK Here are some important words from this reading. Do you understand all of these words? Circle the words you do not understand. Then find the words in the reading. Talk about the meaning of these words with your classmates. If you know all the words, continue to the Question Review.

> balanced diet   legumes
> cereal grains   fiber
> dairy   tofu
> diet

QUESTION REVIEW Go back to the questions on pages 96 and 97. Look at your answers. Work with a partner. Look at your partner's answers too. Are they the same as your answers? Help each other write the correct answers.

PRE-READING 3

FOCUS QUESTION

Skim pages 100 and 101 to find the answer to the question below. Underline the answer in your book. Write the answer below.

- *How does exercise help your body stay healthy?* _____

_____

DETAIL QUESTIONS

LEARNING STRATEGY
☆ Reading selectively

Read "Exercise" on pages 100 and 101. Find the details. Underline the answers in your book. Write the answers below. As you read, write down on your vocabulary tickets any words you do not understand or cannot pronounce.

1. What is muscle tone? _____

_____

2. Finish this sentence: Your body will use more calories when the _____ increases.

3. What is aerobic exercise? _____

_____

4. Give an example of a nonaerobic exercise. _____

_____

5. How much exercise should you get? _____

_____

6. What exercise is best for you? _____

_____

READING 3 ★

Exercise

Exercise is important for a healthy body. Exercise can help your body in two important ways. First, exercise improves **muscle tone**. Your muscles are firmer, stronger, and react more quickly when you exercise. Exercise also increases **metabolic rate**. Metabolic rate is the rate at which energy is used in your body. When the metabolic rate increases, your body uses more calories. For example, if you **jog** or run slowly for 20 minutes, you will use about 180 calories. If you do not consume more calories than your body needs and you jog daily for 20 minutes, you will lose weight.

There are two kinds of exercise, **aerobic exercise** and **nonaerobic exercise**. Aerobic exercise is exercise that increases your heart rate and rate of breathing. When you exercise your heart beats faster, and it pushes more blood through the body. Some examples of aerobic exercises are biking, swimming, jogging, and cross-country skiing. Aerobic exercise is also steady and continuous. For example, when you bike, you move steadily and continuously, you do not stop and start for short periods of time. In nonaerobic exercises, such as stretching, tennis, walking slowly, or weight lifting, the heart rate is not steady and continuous.

Jogging and biking are aerobic activities. Tennis and calisthenics are nonaerobic activities.

How much exercise should you get? It is important to get the right amount of exercise. For the best result exercise aerobically for 15 minutes, three to five times a week. What kind of exercise is best? There are many different kinds of exercises that will help you stay healthy. The most important thing is to do something you enjoy. Exercise will not do you any good if you do not do it!

In this reading you learned why exercise is important for a healthy body. You also learned about the two types of exercise.

SELF-EVALUATION 3

VOCABULARY TICKETS Read the vocabulary tickets with your teacher and the whole class. Are there still some words you do not understand? Write these words in a notebook. With a partner, write some example sentences using these new words. Talk about the meaning of these words with your classmates.

VOCABULARY CHECK

Here are some important words from this reading. Do you understand all of these words? Circle the words you do not understand. Then find the words in the reading. Talk about the meaning of these words with your classmates. If you know all the words, continue to the Question Review.

aerobic exercise
jog
metabolic rate
muscle tone
nonaerobic exercise

QUESTION REVIEW

Go back to the questions on page 100. Look at your answers. Work with a partner. Look at your partner's answers too. Are they the same as your answers? Help each other write the correct answers.

CHAPTER REVIEW

Now that you have completed your reading about nutrition and exercise, go back to pages 93 and 94. Look at your first ideas about nutrition and exercise. Have your ideas changed? What have you learned? Talk about your ideas with the teacher and the whole class.

EXTENSION ACTIVITIES

A. FOOD GROUPS

LEARNING STRATEGY
☆ Grouping

The food you eat helps your body in many different ways. Food can be divided into four basic groups. These four groups give your body different kinds of help. Look at the chart below. Add more foods in each group.

Protein Group	Dairy Group	Fruit and Vegetable Group	Bread and Cereal Group
steak	cheese	bananas	oatmeal

B. MEAL PLANNING

Sit down with three or four of your classmates. One person in your group should be the secretary and record the answers in a notebook. Your group should make a balanced meal plan for two days. Use food from all four food groups each day. Be prepared to share your meal plans and explain why they represent a balanced diet.

C. AEROBIC OR NONAEROBIC EXERCISE?

Sit down with three or four of your classmates. One person in your group should be the secretary and write the answers. Look at the exercise below. Decide if the exercise is aerobic or nonaerobic. Be prepared to explain your answers to the rest of the class.

LEARNING STRATEGIES
☆ Grouping
☆ Taking notes

Exercise	*Aerobic*	*Nonaerobic*
Running	_____	_____
Vacuuming	_____	_____
Downhill skiing	_____	_____
Running in place	_____	_____
Climbing	_____	_____
Handball	_____	_____
Swimming	_____	_____
Rowing	_____	_____
Square dancing	_____	_____
Golf	_____	_____
Jumping rope	_____	_____
Jogging	_____	_____
Weight lifting	_____	_____
Washing windows	_____	_____

GLOSSARY

aerobic exercise Exercise that increases the heart rate and rate of breathing.

balanced diet To eat the right foods in the right amounts.

calories The energy value in food; units to measure the energy that foods give the body.

carbohydrates A class of nutrients; sugars and starches in foods.

cereal grains Grains such as oats, wheat, and rice, used to make cereal.

dairy Related to milk; foods made from milk.

diet The foods a person eats on a regular basis.

essential Very important or necessary.

fats A class of nutrients, such as oils and meat, used for energy storage.

fiber Rough, raw, or coarse food, such as grains and raw fruits and vegetables.

human ecology The study of human beings in the world they inhabit.

jog To run slowly at a steady pace.

legumes Beans or peas.

metabolic rate The rate at which energy is used in the body.

minerals A class of nutrients found in fruits and vegetables, for example. They are important for growth.

muscle tone A healthy, firm condition of the muscle.

nonaerobic exercise Exercise that is not steady and continuous.

nutrients Substances in foods that the body needs in order to be healthy; there are six classes of nutrients.

proteins A class of nutrients found mostly in beans and meat. They are important for growth.

tofu A food made from soybeans that is high in protein.

vitamins Special substances in our foods that keep the body healthy and promote growth.